Lectures on operator theory

Ramanujan Mathematical Society Lectures Notes Series

Vol. 1: Number theory

Vol. 2: The Riemann zeta function and related themes

Vol. 3: Formal language aspects of natural computing

Vol. 4: Commutative algebra and combinatorics

Vol. 5: Convexity in discrete structures

Vol. 6: Number theory and discrete geometry

Vol. 7: Discrete mathematics

Vol. 8: Lectures on operator theory

Vol. 9: Essays on geometric group theory

Vol. 10: Techmüller theory and moduli problems

Ramanujan Mathematical Society

Lecture Notes Series

Volume 8

Lectures on operator theory

Proceedings of the Advances Instructional
School on functional and harmonic analysis,
and the workshop on operator theory,
Indian Statistical Institute,
Bangalore, June/July 2006

Volume editors

T. S. S. R. K. Rao

International Press
www.intlpress.com

Ramanujan Mathematical Society
Lecture Notes Series, Volume 8
Lectures on operator theory

Volume Editor:
T. S. S. R. K. Rao

Published in 2010 by International Press, Somerville, Massachusetts, U.S.A. under license from the Ramanujan Mathematical Society.

This work is also published in the Republic of India, exclusively by the Ramanujan Mathematical Society.

ISBN 978-1-57146-193-3

Typeset using the LaTeX system.
Printed in the United States of America.

Preface

The National Board for Higher Mathematics (NBHM) of India has initiated a program of Advanced Training Schools (ATM) in Mathematics with a view to provide training in various areas of Mathematics to research scholars and young faculty, working in various Indian Universities and Institutes.

An Advanced Instructional School (AIS) on Functional and Harmonic Analysis was held in July 2006 at the Indian Statistical Institute (ISI), Bangalore under the supervision of Dr T. S. S. R. K. Rao and Dr E. K. Narayanan. Lectures covering basic aspects of Functional Analysis and Spectral Theory were given by Dr Rao and Dr B. Bagchi in this program.

As a follow up of this program a week long 'Workshop on Operator Theory' was held at ISI Bangalore in June 2007. Professor G. Godefroy of the University of Paris VI and Professor G. L. Price of the US Naval Academy have lectured in this program on 'Some linear operators and their invariant subsets' and 'An introduction to E-0-semigroups. Professor Godefroy also gave a memorial lecture in the honor of Professor G. Choquet.

This monograph contains a reproduction of the notes prepared by the speakers during these programs. I thank the NBHM for completely funding the above programs including international travel of the speakers and the Ramanujan Mathematical Society for agreeing to bring out the lecture notes in this form. I also thank Professor S. K. Pal, Director of ISI for his encouragement and the Head of the ISI at Bangalore, faculty of ISI, the Indian Institute of Science, co-workers, for the logistic support. Thanks are also due to Mrs Ashalatha and Miss Mohanadevi for their help in typing these notes. The authors thank Professor A. Mangasuli for carefully proof reading these notes. I thank my wife Sireesha for tolerating my busy summer schedules.

T. S. S. R. K. Rao

Contents

Proc. Int. Conf. – Lectures on Operator Theory No. 8, 2009, pp. 1–20.

Spectral Theory in Banach Algebras

Bhaskar Bagchi

Statistics and Mathematics Unit, Indian Statistical Institute, R. V. College Post,
Bangalore 560 059, India
e-mail: bbagchi@isibang.ac.in

1. Motivation

Let V be a vector space over the field $\mathbb{C}$ of complex numbers and let $T : V \to V$ be a linear operator. First suppose V is finite dimensional. Recall that, in this case, the notion of *eigenvalues* of T plays a crucial role in the analysis of T, particularly when V is equipped with an inner product and T is self-adjoint (more generally, normal). A scalar $\lambda \in \mathbb{C}$ is an eigenvalue of T if there is a vector $x \neq 0$ in V such that $Tx = \lambda x$. (Any such x is a corresponding eigenvector, and the set of all such x, together with 0, is the eigenspace of λ.) In other words, λ is an eigenvalue if $(T - \lambda I)$ has a non-trivial kernel (and $\ker(T - \lambda I)$ is the eigenspace), equivalently, when $T - \lambda I$ fails to be one-one. But, in the finite dimensional situation, a linear operator (such as $T - \lambda I$) is one-one if and only if it is onto. Thus, when V is finite dimensional and $T : V \to V$ is linear, $\lambda \in \mathbb{C}$ is an eigenvalue of T iff $T - \lambda I$ is *not* invertible. In this case the set $\sigma(T)$ consisting of all the eigenvalues of T is called the spectrum of T. Thus,

$$\sigma(T) = \{\lambda \in \mathbb{C} : T - \lambda I \text{ is not invertible}\}.$$

A crucial starting point in linear algebra is the fact that every linear operator has at least one eigenvalue: $\sigma(T) \neq \phi$.

Exercise. Can you give an example $T : V \to V$ which has only one eigenvalue? An example $T : V \to V$ on a f.d. real vector space with no eigenvalues?

How do we prove this fact? The usual proof you see in a linear algebra course rests on the notion of determinants. Here, we just recall the following facts: (i) $\det(T)$ is a complex number attached to T such that $\det(T) \neq 0$ iff T is invertible and (ii) $\lambda \mapsto \det(T - \lambda I)$ is a polynomial of degree $n := \dim V$.

Thus, $\sigma(T)$ is precisely the set of zeros of this polynomial. Hence, $\sigma(T)$ is finite (in fact, T has at most n eigenvalues) and non-empty (by the "fundamental theorem of algebra": $\mathbb{C}$ is algebraically closed.) This is why we have chosen the field $\mathbb{C}$ rather than $\mathbb{R}$.

(The usual definition of $\det(T)$ is rather messy. First one gives a rather strange looking definition of $\det(A)$ for an $n \times n$ matrix A and investigates its properties. Next, one associates a matrix A with the operator T, depending on the choice of an ordered basis of V, and defines the determinant of the operator T to be the determinant of A. It turns out that all the matrices A associated with T (corresponding to all choices of the ordered basis of V) are similar and hence have the same determinant. Thus $\det(T)$ is well-defined. This is a messy and round about definition. The satisfactory and direct definition of $\det(T)$ comes from the theory of wedge product and exterior algebra. If you have not seen this, then rush to the library and pick up a book on multilinear algebra!)

Exercise. Must we use determinants to prove that $\sigma(T)$ is non-empty and finite?

No!, one can try to prove (without using determinants!) that $\lambda \mapsto (T - \lambda I)^{-1}$ is a (operator-valued!) rational function which is bounded at infinity (in the sense that if T is represented by a matrix, then all the n^2 co-ordinates-entries-of the matrix of $(T - \lambda I)^{-1}$ are rational functions bounded at ∞.) So if it had no "poles" then Liouville's theorem would be contradicted. This is the kind of argument that we shall generalize to the infinite dimensional setting.

(Note that the space $\mathbf{W}$ of all linear operators from V to V is a $\mathbb{C}$-vector space in its own right. Given a $\mathbb{C}$-vector space $\mathbf{W}$ and a $\mathbf{W}$-valued function on a domain in $\mathbb{C}$, does it make unambiguous sense to say that this is a rational function or must we refer to a chosen co-ordinate system (basis) to make sense of this notion? A serious perusal of this question again leads to symmetric tensor products and multilinear algebra)

2. Bounded linear operators on a Banach space and their spectra

We wish to develop an analogue of all the above in the context of infinite dimensional vector spaces. To avoid pathologies and to get a good generalization, it seems that we must introduce a topology (i.e. notion of convergence) on our vector space. To be of any use to us, it must be compatible with the vector space structure, i.e. scalar multiplication and vector addition must be continuous with respect to this topology. This enables us to restrict attention to continuous linear operators. (This is not to say that discontinuous operators can't be interesting, quite the contrary. They are just that much more difficult to analyse: there is no hope of a general theory. Moreover, the interesting discontinuous operators that occur "in nature" are usually not everywhere defined on the vector spaces.)

The simplest way to introduce a compatible topology on a vector space V is to give ourselves a norm $\|\cdot\|$ on V. Recall that this is a function $\|\cdot\| : V \to [0, \infty)$ such that (i) $\|x\| = 0$ iff $x = 0$ (non-degenerate), (ii) $\|\lambda x\| = |\lambda| \cdot \|x\|$ for $\lambda \in \mathbb{C}$, $x \in V$(homogeneous) and (iii) $\|x + y\| \leq \|x\| + \|y\|$ for all $x, y \in V$ (sub-additive).

Given such a norm, one defines the distance between two vectors in V to be the norm of their difference. This leads to a natural extension of the notions of Cauchy and convergent sequences. It is easy to see that the resulting topology is compatible with the vector space structure.

If V is a normed vector space, then for any fixed vector x_0 and any scalar $\lambda_0 \neq 0$, the map $x \mapsto x_0 + \lambda_0 x$ from V to V is easily seen to be a homeomorphism from V onto itself. Using these homeomorphisms, one may move any closed ball in V onto any other closed ball homeomorphically. Thus, all closed balls in V "look alike". Further, these homeos behave nicely with respect to linear operators. Using this observation, it is easy to see that if $T : V_1 \to V_2$ is a linear operator between normed vector spaces, then the following statements are equivalent:- (i) T is continuous, (ii) T is bounded on some closed ball, (iii) T is bounded on all closed balls.

Thus, $T : V_1 \to V_2$ is continuous iff it is *bounded* in the sense that $\|T\| := \sup_{x \in V \,:\, \|x\| \leq 1} \|Tx\| < \infty$.

Following tradition, we shall use the word "bounded" rather than "continuous" in the context of linear operators between normed vector spaces- even though they mean the same thing.

Exercise. (i) If V_1, V_2 are normed vector spaces, then the space $\mathcal{L}(V_1, V_2)$ of all bounded linear operators from V_1 to V_2 is a normed vector space, with the "operator norm" defined above. (ii) If $\mathbf{X}$ is any metric space, then let $C_b(X)$ denote the vector space of all bounded and continuous complex-valued functions on X, equipped with the "sup-norm": $\|f\|_\infty = \sup_{x \in X} |f(x)|$. Then $C_b(X)$ is a normed vector space.

To make things simple, we shall assume that all normed vector spaces we consider are complete, in the sense that all Cauchy sequences in the space converge. A complete normed vector space is called a *Banach space*.

Exercise. (i) In the study of linear operators, the assumption of completeness is *not* a serious one. This is because, as shown in the next part of the exercise, any normed vector space V can be embedded as a dense linear subspace (with induced structure) of a Banach space $\bar{V}$. Now, if $T : V_1 \to V_2$ is a bounded linear operator between normed vector spaces, then T extends uniquely to a bounded linear operator $\bar{T} : \bar{V_1} \to \bar{V_2}$, and we may identify T with $\bar{T}$ for many purposes. (ii) If V_1, V_2 are normed spaces, then $\mathcal{L}(V_1, V_2)$ is Banach provided V_2 is Banach (even if V_1 is not). In particular, for any normed vector space V, its *dual space* $V^* := \mathcal{L}(V, \mathbb{C})$ is a Banach space. Define $\wedge : V \to V^{**}$ by $\hat{x}(f) = f(x), x \in V, f \in V^*$. Then check that $\wedge$ is a linear isometry from V into (usually not onto) V^{**}. Therefore, we may identify V with its image $\hat{V} \subseteq V^{**}$. Now define $\bar{V}$ to be the closure in V^{**} of $\hat{V}$. Being a closed linear subspace of a Banach space, $\bar{V}$ is a Banach space in which (a copy of) V sits as a dense linear subspace. [To verify that $\wedge$ is an isometry, one needs the Hahn-Banach Theorem.]

So, from now on, all vector spaces (not necessarily finite dimensional) are Banach spaces.

Now we wish to define the spectrum $\sigma(T)$ of a bounded linear operator, T on a Banach space V. Recall that, even in the finite dimensional situation, this definition depends on the notion of invertibility. So when should we call a bounded linear operator $T : V \to V$ invertible? Is it enough to insist T is invertible as a function, or should we also insist that its inverse is bounded (i.e. that T is a homeomorphism)? Fortunately, the open mapping theorem guarantees that when a bounded linear operator on a Banach space is invertible as a set map, the inverse is automatically bounded.

Now, as before, we define

$$\sigma(T) = \{\lambda \in \mathbb{C} : T - \lambda I \text{ is not invertible}\}.$$

As before, we may define an eigenvalue λ of T to be a complex number such that $\ker(T - \lambda I) \neq \{0\}$. But $\sigma(T)$ may now have points which are not eigenvalues! This is because a 1-1 linear map on an infinite dimensional Banach space (to itself) need not be onto, and vice versa.

Exercise. (i) Given $T : V \to V$, recall that the adjoint $T^* : V^* \to V^*$ is defined by

$$(T^* f)(x) = f(Tx), x \in V, \ f \in V^*.$$

Show that

$$\sigma(T^*) = \overline{\sigma(T)} := \{\bar{\lambda} : \lambda \in \sigma(T)\}.$$

(ii) Consider the Banach space ℓ^2 whose elements are the absolutely square summable sequences $x = \{x_n : n = 1, 2, \dots\}$ with norm given by $\|x\|^2 = \sum_{n=1}^{\infty} |x_n|^2$. (Can you show that this is actually a Banach space?) Define the operator $S : \ell^2 \to \ell^2$ by $S(x_1, x_2, \dots) = (0, x_1, x_2, \dots)$. S is called the (forward) unilateral shift. Note that S is a linear isometry. In consequence, $\|S\| = 1$, so S is a bounded operator. Prove that S has no eigenvalue! Also show that every complex number w with $|w| < 1$ is an eigenvalue of S^*. Thus $\sigma(S^*)$ and hence also $\sigma(S)$, contains the open unit disc. Actually, as we shall soon see, $\sigma(S)$ is a closed set, hence contains the closed unit disc. Further, as $\|S\| = 1$, it will transpire that $\sigma(S)$ is contained in the closed unit disc (with centre 0) in $\mathbb{C}$. Thus, in fact, $\sigma(S)$ is the closed unit disc, though S has no eigenvalues.

Example. Let $\mathcal{H}$ denote the Banach space whose elements are all infinite sequences $\{x_n\}$ of complex numbers such that $\sum_{n=1}^{\infty} \frac{|x_n|^2}{n^2} < \infty$. Equip it with coordinate-wise addition and scalar multiplication. Also, norm it by: $\|x\|^2 = \sum_{n=1}^{\infty} |\frac{x_n}{n}|^2$, $x \in \mathcal{H}$. Define a linear operator $T : \mathcal{H} \to \mathcal{H}$ by $Tx = y$ where $y_m = \sum_{n=1}^{\infty} \frac{\{\frac{n}{m}\}x_n}{n^2}$, $m = 1, 2, \dots$ (Here $\{\cdot\}$ is the fractional part function). Show that T is indeed a bounded linear operator!

Question. Is this operator $T : \mathcal{H} \to \mathcal{H}$ one-one? In other words, is 0 an eigenvalue of T? (It is easy to see that $0 \in \sigma(T)$). Surprisingly, nobody knows the answer! Indeed, a recent result may be restated as saying that the Riemann Hypothesis (one of the greatest unsolved problems in the whole of Mathematics) amounts to the assertion that 0 is not an eigenvalue of this operator (see [1]). So, just because one can glibly quote a few definitions, one should not imagine that the underlying notions are simple and totally under control!

Coming back to the general bounded linear operator $T : V \to V$ on a Banach space V. Let's show that $\sigma(T)$ is a bounded subset of $\mathbb{C}$. Indeed, we shall see that $\sigma(T) \subseteq \{\lambda \in \mathbb{C} : |\lambda| \leq \|T\|\}$, the closed disc centered at 0 of radius $\|T\|$.

Now, the spectrum is covariant under scaling. More precisely, for any non-zero scalar c, $\sigma(cT) = c\sigma(T) := \{c\lambda : \lambda \in \sigma(T)\}$. Using this idea, we find that, to prove the above inclusion, it suffices to prove the following assertion (which, on the face of it, looks like a very special case of the inclusion):-

Assertion. If $\|T\| < 1$ then $1 \notin \sigma(T)$. In other words, if $\|T\| < 1$ then $I - T$ is invertible. The idea of the proof is very simple: that of a geometric series. We all know that if z is a complex number with $|z| < 1$, then the series $\sum_{n=0}^{\infty} z^n$ converges absolutely to $(1 - z)^{-1}$. This simple thing can be jacked up here! Just show that $(I - T)^{-1}$ is given by the series $\sum_{n=0}^{\infty} T^n$.

Of course, one needs to be careful. Use the fact that operator norm is sub-additive and submultiplicative ($\|T_1 + T_2\| \leq \|T_1\| + \|T_2\|$, $\|T_1 T_2\| \leq \|T_1\| \cdot \|T_2\|$) to establish that for each vector $x \in V$, the series $\sum_{n=0}^{\infty} T^n x$ is Cauchy in V (i.e., the sequence of partial sums is Cauchy), hence it converges to a vector in V. (For the first time, we are using our assumption that V is a Banach space.) So one can define $S : V \to V$ by $Sx = \sum_{n=0}^{\infty} T^n x,\ x \in V$. Now it should be easy to conclude that S is a bounded linear operator, and an essentially algebraic manipulation (as with the geometric series of scalars) will show that $S(I-T) = I = (I-T)S$. Thus S is the inverse of $I-T$: $I-T$ is invertible.

Writing down all the details of this proof carefully, we have now proved:

Lemma 1. *If $T : V \to V$ is a bounded linear operator on a Banach space V, then its spectrum $\sigma(T)$ is bounded. Indeed, $\sigma(T)$ is contained in the closed disc of radius $\|T\|$.*

The next thing we wish to prove is that $\sigma(T)$ is closed. In other words, $\mathbb{C}\backslash\sigma(T)$ is open. So we need to show that if $\lambda_0 \in \mathbb{C}\backslash\sigma(T)$ then every scalar λ sufficiently close to λ_0 (in symbol, $\lambda \approx \lambda_0$) is in $\mathbb{C}\backslash\sigma(T)$.

So take $\lambda \approx \lambda_0$. (Exactly how close is sufficiently close, we shall decide later.) Then,

$$\begin{aligned} \lambda I - T &= (\lambda_0 I - T) - (\lambda_0 - \lambda)I \\ &= (\lambda_0 I - T)(I - (\lambda_0 - \lambda)(\lambda_0 I - T)^{-1}) \end{aligned}$$

Now, $\lambda_0 I - T$ is invertible. So, to conclude that $\lambda I - T$ is also invertible, it suffices to show that the other factor is invertible. But, in view of the "geometric series" trick, this will be so provided $\|(\lambda_0 - \lambda)(\lambda_0 I - T)^{-1}\| < 1$ i.e., whenever $|\lambda - \lambda_0| < \varepsilon := \frac{1}{\|(\lambda_0 I - T)^{-1}\|}$. We have proved that the spectrum of any bounded linear operator on a Banach space is closed. It is also bounded. As closed and bounded subsets of $\mathbb{C}$ are compact, it follows that

Theorem. *If $T \in \mathcal{L}(V) := \mathcal{L}(V, V)$ for a Banach space V, then $\sigma(T)$ is compact.*

Notice that "compactness" is the correct topological generalization of "finiteness" – which is what makes this notion so useful. For instance, any real-valued function on a compact set achieves its maximum provided the function is continuous. Thus, the above theorem is the natural generalization of the fact that the spectrum of a finite-dimensional operator is finite.

Example. Let $\mathcal{H}$ be a separable Hilbert space (i.e. a Banach space whose norm is defined by an inner product). Fix an orthonormal basis $\{e_n : n = 1, 2, \dots\}$ of $\mathcal{H}$. (For instance, we may take $\mathcal{H} = \ell^2$ and e_n, the standard basic vector whose nth entry is $= 1$ and other entries $= 0$). Now, take any *bounded* sequence $\{\lambda_n : n \geq 1\}$ of complex numbers. Show that there is a (unique) *bounded linear operator* $T : \mathcal{H} \to \mathcal{H}$ such that $Te_n = \lambda_n e_n$ for $n = 1, 2, \dots$ (For obvious reasons, such an operator is called a *diagonal operator.*) Prove that $\sigma(T)$ is the closure in $\mathbb{C}$ of the sequence $\{\lambda_n : n = 1, 2, \dots\}$. Now take any non-empty compact set $K \subset \mathbb{C}$. Then we can find a (bounded) sequence in K whose closure is K. Therefore, every non-empty compact subset of $\mathbb{C}$ occurs as the spectrum of a bounded linear operator.

Exercise. Let G be the set of all invertible bounded linear operators from a Banach space V to itself. Show that G is an open subset of $\mathcal{L}(V)$. (Hint:- You need to show that for any $T_0 \in G$, $\exists\, \epsilon > 0$ such that for $T \in \mathcal{L}(V)$, $\|T - T_0\| < \epsilon \Rightarrow T \in G$. Show that $\epsilon = \frac{1}{\|T_0^{-1}\|}$ works.)

In the above discussion of the spectrum of operators T, we never used the fact that T was an operator! All we used was that operators can be added and multiplied, and they carry a norm which behaves nicely with respect to these algebraic operations. So it may be fruitful in the study of spectra to forget altogether that we are talking about operators! (It helps to forget irrelevant information - this is what the word "abstraction" really means.) This leads us to define a *Banach Algebra* to be a Banach space which carries the structure of an algebra with a multiplicative unit e such that the norm is sub-multiplicative and satisfies $\|e\| = 1$. (Recall that an algebra is a vector space which is also a ring, such that multiplication in the ring and scalar multiplication are related by the requirement $\lambda(xy) = (\lambda x)y = x(\lambda y)$ for scalars λ and vectors x, y).

Thus, for any Banach space V, $\mathcal{L}(V)$, with usual operations and operator norm, is our prime example of Banach algebras. A second example worth keeping in mind is the Banach algebra $C(K)$ of continuous complex-valued functions on a compact Hausdorff space K, with supremum norm. Whatever we have done with operators generalizes readily to vectors in Banach algebras. Thus, given a Banach algebra A, and vector $x \in A$, the spectrum $\sigma(x)$ is by definition the subset $\{\lambda \in \mathbb{C} : x - \lambda e \text{ is not invertible}\}$ of the complex plane.

Exactly as before, we have:

Theorem. *For any vector x in a Banach algebra A, $\sigma(x)$ is a compact subset of $\mathbb{C}$.*

Example. For $f \in C(K)$, $\sigma(f)$ is nothing but the range of $f : \sigma(f) = f(K)$.

Now we proceed to show that for any Banach algebra A, and any vector $x \in A$, $\sigma(x) \neq \phi$. As already indicated, this depends on the analytic properties of the map $\lambda \mapsto (x - \lambda e)^{-1}$ from $\mathbb{C}\backslash\sigma(x)$ into A. But what do we mean by the analyticity of such a map?

More generally, if V is a Banach space, Ω an open subset of $\mathbb{C}$ and $f : \Omega \to V$ is a function, what should we mean by the statement that f is analytic? Of course, the definition should be such that most of the results about the usual (complex valued) analytic functions go through.

Three competing definitions suggest themselves:

1) The Weierstrass definition: $f : \Omega \to V$ is analytic if f is locally given by power series (with coefficients from V). That is, for each $\lambda_0 \in \Omega$, there is a neighbourhood $\mathbf{U}$ of λ_0 such that $\mathbf{U} \subseteq \Omega$ and $f(\lambda) = \sum_{n=0}^{\infty} a_n(\lambda - \lambda_0)^n$, $\lambda \in \mathbf{U}$, and the convergence is in the norm of V.

2) The strong definition: $f : \Omega \to V$ is analytic if for every $\lambda_0 \in \Omega$, $f'(\lambda_0) := \lim\limits_{\lambda\to\lambda_0} \frac{f(\lambda)-f(\lambda_0)}{\lambda-\lambda_0}$ exists. (Here again the limit is with respect to the norm in V.)
3) The weak definition: f is analytic if, for every bounded linear functional $\varphi \in V^*$, the complex valued function $\varphi \circ f$ on Ω is analytic in the usual sense.

Now, a magical thing happens. All three definitions of analyticity are actually equivalent, so that we have a single natural definition of analyticity of Banach space valued functions on planar domains. Some of these implications are easy to see. The simplest is strong analyticity implies weak analyticity: if f is strongly analytic, $\lim\limits_{\lambda\to\lambda_0} \frac{f(\lambda)-f(\lambda_0)}{\lambda-\lambda_0}$ exists for each $\lambda_0 \in \Omega$, and hence, for each $\varphi \in V^*$,

$$\lim_{\lambda\to\lambda_0} \frac{\varphi\circ f(\lambda)-\varphi\circ f(\lambda_0)}{\lambda-\lambda_0} = \lim_{\lambda\to\lambda_0} \varphi\left(\frac{f(\lambda)-f(\lambda_0)}{\lambda-\lambda_0}\right)$$

exists, using linearity and continuity of φ. The implication, Weierstrass analyticity $\Rightarrow$ strong analyticity, is also easy:

Remark. Let $\sum_{n=0}^{\infty} a_n z^n$ be a power series in the complex variable z, where the coefficients a_n are from a Banach space V. Such a power series shares all the usual elementary properties of the usual power series with complex coefficients; with the same proof. Thus, there is a radius of convergence $0 \le R \le \infty$, given by the usual formula, $R = \frac{1}{\limsup\limits_{n\to\infty} \|a_n\|^{1/n}}$ such that the power series converges in the open disc in $\mathbb{C}$ with centre 0 and radius R, diverges in the exterior (complement of closure) of this disc, the convergence is absolute (i.e, $\sum \|a_n\|\,|z|^n < \infty$) in the interior of this disc, and the power series can be differentiated term by term in the interior of this disc. Thus Weierstrass analyticity implies strong analyticity.

Thus we have the implications Weierstrass analytic $\Rightarrow$ strongly analytic $\Rightarrow$ weakly analytic.

To prove the rest of the implications (to establish the equivalence of the three definitions), we need to develop an integration theory for Banach space valued functions. Luckily, we need to do this only for continuous functions.

Recall that a subset X of a vector space V is said to be *convex* if the line segment joining any two points in X is contained in X (i.e, $x_1, x_2 \in X$, $0 \le t \le 1 \Rightarrow tx_1 + (1-t)x_2 \in X$). For any set $Y \subseteq V$ the *convex hull* of Y, denoted conv(Y), is the intersection of all convex subsets of V containing Y. conv(Y) is the "smallest" convex subset of V which contains Y.

Lemma. *Let K be a compact subset of the Banach space V. Then the closed convex hull of K (= the closure of its convex hull) is compact.*

Sketch of Proof. Since V is a complete metric space, a subset of V is compact if and only if it is closed and totally bounded (a set is totally bounded if for each $\epsilon > 0$, it can be covered by finitely many open balls of radius ϵ). So it is enough to show: K totally bounded $\Rightarrow$ conv(K) totally bounded. Fix $\epsilon > 0$. Since K is totally bounded, there is a finite set $E \subseteq V$ such that the open balls of radius $\frac{\epsilon}{2}$ with centres in E cover K. Let C denote the convex hull of E. Being the convex hull of a finite set, C is compact (why?). Therefore, there is a finite set $F \subseteq V$ such that the open balls with radius $\frac{\epsilon}{2}$ and centres in F cover C. Now argue that the open balls with radius ϵ and centres in the finite set F cover conv(K). Since $\epsilon > 0$ was arbitrary, this shows that conv(K) is totally bounded, hence its closure is compact.

Now we are ready to introduce vector-valued integration.

Theorem/Definition. *Let Q be a compact Hausdorff space, μ a Borel probability on Q, V a Banach space and $f : Q \to V$ a continuous function. Then there is a unique vector $v_0 \in V$ such that $\varphi(v_0) = \int_Q (\varphi \circ f) d\mu \; \forall \; \varphi \in V^*$. Further, $v_0 \in \overline{\text{conv } f(Q)}$. We define, $\int_Q f d\mu = v_0$.*

(Thus, by definition, we have

$$\varphi\left(\int_Q f d\mu\right) = \int_Q \varphi \circ f \, d\mu, \; \varphi \in V^*,$$

as we ought to have!)

Proof. Uniqueness is clear since if we had two such vectors v_0, ω_0 then $\varphi(v_0) = \varphi(\omega_0) \; \forall \; \varphi \in V^*$. Since by Hahn-Banach theorem, V^* separates points of V, this would force $v_0 = \omega_0$.

So it suffices to prove existence of v_0. To each $\varphi \in V^*$, we associate the set

$$E_\varphi = \{v \in \overline{\text{conv } (f(Q))} : \phi(v) = \int_Q \varphi \circ f d\mu\}.$$

By continuity of φ, this is a closed subset of $\overline{\text{conv } f(Q)}$. Since Q is compact and f continuous, it follows that $f(Q)$ is compact. Hence, by the lemma, $\overline{\text{conv } (f(Q))}$ is compact. Being a closed subset of a compact set, E_φ is compact for every $\varphi \in V^*$.

We need to show that the intersection $\bigcap_{\varphi \in V^*} E_\varphi$ is non-empty: then any v_0 in the intersection satisfies the requirement of the theorem. Since each E_φ is compact, their intersection would be non-empty if the family $\{E_\varphi : \varphi \in V^*\}$ had finite intersection property.

Thus, it suffices to show that if $\varphi_1, \varphi_2, \ldots, \varphi_n$ are finitely many bounded linear functionals on V then the intersection $E_{\varphi_1} \bigcap \ldots \bigcap E_{\varphi_n}$ is non-empty.

Let $L = (\varphi_1, \ldots, \varphi_n) : V \to \mathbb{C}^n$. Thus L is a bounded linear operator and hence $K := L(f(Q))$ is a compact subset of $\mathbb{C}^n$. Therefore, conv(K) is a compact convex subset of $\mathbb{C}^n$ (the convex hull of a compact set in a finite dimensional Banach space is automatically compact - why?). Since L is linear, we have conv$(K) = $ conv$(L(f(Q))) = L($ conv $(f(Q))) \subseteq L(\overline{\text{conv} f(Q)})$.

Therefore, to complete the proof, it suffices to show that the vector $z = (z_1, \ldots, z_n) \in \mathbb{C}^n$ defined by $z_i = \int_Q \varphi_i \circ f d\mu, \; 1 \leq i \leq n$, is in conv$(K)$. Then $z = L(v)$ for some $v \in \overline{\text{conv} f(Q)}$ and - if you dis-entangle the definitions- you see that $v \in E_{\varphi_1} \bigcap \ldots \bigcap E_{\varphi_n}$.

Suppose, on the contrary, the vector $z \in \mathbb{C}^n$ does not belong to the compact convex set conv$(K) \subseteq \mathbb{C}^n$. Then there is a hyperplane in $\mathbb{C}^n$ which strictly separates z from

$K : z$ lies on one side (one of the two open half-spaces determined by the hyperplane) and K lies on the other side. That is, $\exists\ \varphi \in (\mathbb{C}^n)^*$ such that

$$Re\ \varphi(z) < Re\ \varphi(\omega),\ \forall\ \omega \in K. \qquad (*)$$

But $\varphi \in (\mathbb{C}^n)^*$ is given by $\varphi(v) = \sum_{i=1}^n c_i v_i,\ v \in \mathbb{C}^n$, for some scalars $c_1, \dots c_n \in \mathbb{C}$. Thus, $(*)$ means that

$$\int_Q Re\left(\left(\sum_{i=1}^n c_i\varphi_i\right)\circ f\right) d\mu < Re\left(\left(\sum_{i=1}^n c_i\varphi_i\right)\circ f\right)(q),\ \ \forall\, q \in Q.$$

That is, if $g : Q \to \mathbb{R}$ is the continuous function defined by $g = Re\left(\left(\sum_{i=1}^n c_i\varphi_i\right)\circ f\right)$ then we have $\int_Q g d\mu < g(q)\ \forall\, q \in Q$. This is absurd since μ is a probability measure!

Remark. Since any complex measure μ on Q is a linear combination of four probability measures, it now follows that $\int_Q f d\mu$ may be defined as above for any complex measure μ. It will continue to satisfy $\varphi\left(\int_Q f d\mu\right) = \int_Q \varphi\circ f d\mu$ for all $\varphi \in V^*$. However, if μ is not a probability measure then $\int_Q f d\mu$ need not belong to $\overline{\text{conv} f(Q)}$.

In particular, if $\gamma : [0, 1] \to \mathbb{C}$ is a piecewise smooth path, then the traditional *contour integral* $\int_\gamma f dz$ may be viewed as an integration with respect to a complex measure on range(γ). Alternatively, it may be viewed as the integral of the vector-valued continuous function $f o \gamma$ on $[0, 1]$ with respect to the complex measure $\gamma'(t)dt$ on $[0, 1]$:

$$\int_\gamma f dz := \int_0^1 (f\circ\gamma)(t)\gamma'(t)dt.$$

There remains one little hitch in applying this integration theory to weakly analytic functions: we have defined the integral only for vector-valued *continuous* functions. Therefore, to talk of contour integrals of weakly analytic vector-valued functions, we must first show that they are automatically continuous!

Lemma. *Let Ω be a domain in $\mathbb{C}$, V a Banach space, and $f : \Omega \to V$ be weakly analytic. Then f is continuous.*

Proof. Fix a point $\lambda_0 \in \Omega$ and let $\mathbf{U} \subseteq \Omega$ be a compact neighbourhood of λ_0. To prove continuity of f at λ_0, it suffices to show that the function $g : \Omega\backslash\{\lambda_0\} \to V$, defined by $g(\lambda) = \frac{f(\lambda)-f(\lambda_0)}{\lambda-\lambda_0}$, is bounded on $\mathbf{U}\backslash\{\lambda_0\}$. Then there is a constant $c > 0$ such that $\|f(\lambda) - f(\lambda_0)\| \le c|\lambda - \lambda_0|$ for $\lambda \in \mathbf{U}$, whence f is continuous at $\lambda_0 \in \Omega$.

Recall that a similar result is true for complex-valued analytic functions. Since for each $\varphi \in V^*, \varphi\circ f$ is a complex valued analytic function on Ω, it follows that $\left\{\varphi\left(\frac{f(\lambda)-f(\lambda_0)}{\lambda-\lambda_0}\right) : \lambda \in \mathbf{U}\backslash\{\lambda_0\}\right\} = \left\{\frac{\varphi_0 f(\lambda)-\varphi_0 f(\lambda_0)}{\lambda-\lambda_0} : \lambda \in \mathbf{U}\backslash\{\lambda_0\}\right\}$ is bounded.

Consider the set of vectors $X = \left\{\frac{f(\lambda)-f(\lambda_0)}{\lambda-\lambda_0} : \lambda \in \mathbf{U}\backslash\{\lambda_0\}\right\}$.

The above consideration shows that X is a weakly bounded set of vectors in the sense that $\{\varphi(v) : v \in X\}$ is bounded for each $\varphi \in V^*$. That is, letting $\hat{v} \in V^{**}$ denote the isometric copy of $v \in X$ in V^{**} $(\hat{v}(\varphi) = \varphi(v),\ \varphi \in V^*)$, this means the image

$\hat{X} = \{\hat{v} : v \in X\}$ of X under the linear isometry $\wedge$ is pointwise bounded. Since V^* is a Banach space, the uniform boundedness principle implies that $\hat{X}$ is bounded in norm. Since $\hat{X}$ is the image of X under an isometry, it follows that X is bounded in norm, so we are done.

Remark. Along the way, we have shown that any weakly bounded set of vectors in a Banach space is actually norm-bounded.

Now that we have developed contour integration for continuous functions, and established continuity of weakly analytic functions, we can talk of contour integrals of the latter. It is easy now to establish Cauchy's fundamental theorem for weakly analytic vector-valued functions:

Theorem. *Let $f : \Omega \to V$ be a Banach space valued weakly analytic function on a simply connected planar domain Ω. Then, for any simple closed curve γ in Ω, $\int_\gamma f = 0$.*

Proof. Let $v = \int_\gamma f \in V$. Since $\varphi o f$ is a complex-valued analytic function on Ω for each fixed $\varphi \in V^*$, the usual Cauchy theorem shows that

$$\varphi(v) = \varphi\left(\int_\gamma f\right) = \int_\gamma \varphi \circ f = 0.$$

Since $\varphi(v) = 0 \; \forall \; \varphi \in V^*$, Hahn-Banach implies $v = 0$.

Now, as in ordinary complex analysis, one can prove Cauchy's integral formula for weakly analytic functions, and, expanding the integrand in power series, and integrating term by term, see that weakly analytic functions are Weierstrass analytic. Details are left to you.

Thus we have seen:

Weierstrass analytic $\Rightarrow$ strongly analytic $\Rightarrow$ weakly analytic $\Rightarrow$ Weierstrass analytic.

Hence all three definitions of analytic function into a Banach space are equivalent. So we shall use the adjective "analytic" for such functions without further ado.

It is now easy to extend most elementary results in complex analysis to V-valued analytic functions. For instance:

Liouville's Theorem. *Let $f : \mathbb{C} \to V$ be a Banach space valued entire function. If f is bounded then f must be constant.*

Proof. Suppose $\|f(\lambda)\| \leq c \, \forall \, \lambda \in \mathbb{C}$. Then for each $\varphi \in V^*$, $|\varphi \circ f(\lambda)| \leq c \cdot \|\varphi\| \, \forall \, \lambda \in \mathbb{C}$. Since $\varphi \circ f$ is entire, it follows that it is a constant. Now if f is not constant then $f(\lambda_1) \neq f(\lambda_2)$ for some $\lambda_1, \lambda_2 \in \mathbb{C}$. Hence by Hahn-Banach, we can find $\varphi \in V^*$, such that $\varphi(f(\lambda_1)) \neq \varphi(f(\lambda_2))$. This contradicts constancy of $\varphi \circ f$.

Theorem. *For any Banach algebra A, and vector $x \in A$, the spectrum $\sigma(x)$ is a non-empty, compact set.*

Proof. We have already proved that $\sigma(x)$ is compact. Suppose, if possible, that $\sigma(x)$ is empty. Then $x - \lambda e$ is invertible for all $\lambda \in \mathbb{C}$. Therefore we can define a function

$f : \mathbb{C} \to A$ by $f(\lambda) = (x - \lambda e)^{-1}$. We claim that f is an entire function. To see this, fix any $\lambda_0 \in \mathbb{C}$. Then, for $|\lambda - \lambda_0| < \frac{1}{\|(\lambda_0 e - x)^{-1}\|}$, we have the following valid computation:

$$\begin{aligned} f(\lambda) &= (x - \lambda e)^{-1} = ((x - \lambda_0 e) - (\lambda - \lambda_0)e)^{-1} \\ &= (x - \lambda_0 e)^{-1}(e - (\lambda - \lambda_0)(x - \lambda_0 e)^{-1})^{-1} \\ &= (x - \lambda_0 e)^{-1} \sum_{n=0}^{\infty} (\lambda - \lambda_0)^n (x - \lambda_0 e)^{-n} \\ &= \sum_{n=0}^{\infty} (\lambda - \lambda_0)^n (x - \lambda_0 e)^{-(n+1)} \\ &= \sum_{n=0}^{\infty} a_n (\lambda - \lambda_0)^n \end{aligned}$$

Where $a_n = (x - \lambda_0 e)^{-(n+1)} \in A$ for $n = 0, 1, 2, \ldots$ Thus f is given locally by a convergent power series around any $\lambda_0 \in \mathbb{C}$. Thus f is entire. We now claim that f is norm-bounded. For this, notice that for $|\lambda|$ large, we have the valid computation

$$\begin{aligned} f(\lambda) &= -(\lambda e - x)^{-1} = -\lambda^{-1}(e - \lambda^{-1}x)^{-1} \\ &= -\lambda^{-1} \sum_{n=0}^{\infty} \lambda^{-n} x^n \\ &= -\sum_{n=0}^{\infty} \lambda^{-n-1} x^n, \text{ (valid for } |\lambda| > \|x\|) \end{aligned}$$

Hence $\|f(\lambda)\| \leq \sum_{n=0}^{\infty} |\lambda|^{-n-1} \|x\|^n = (1 - |\lambda|^{-1}\|x\|)^{-1} \cdot |\lambda|^{-1} \to 0$ as $|\lambda| \to \infty$.

Thus f is bounded; indeed $f(\lambda) \to 0$ as $\lambda \to \infty$. Hence, by Liouville's Theorem, f is constant. Since $f \to 0$ near infinity, it follows that $f \equiv 0$.

In particular, $x^{-1} = f(0) = 0$. Multiplying by x, we get $e = 0$: a contradiction.

Remark. This proof goes through, in particular, for the finite dimensional Banach algebra $\mathbb{C}^{n \times n}$ of $n \times n$ complex matrices. As promised, it does not require the determinant.

As an immediate Corollary of the above theorem we have:

Gelfand-Mazur Theorem. *If A is a Banach algebra which is also a field (i.e. every non-zero element of A is invertible) then A is isometrically isomorphic to the field $\mathbb{C}$ of complex numbers.*

In short, $\mathbb{C}$ (with usual modulus as norm) is the only Banach algebra which is a field.

Proof. Consider the map $\lambda \mapsto \lambda e$ from $\mathbb{C}$ into A. Since $\|e\| = 1$, this is a linear isometry. So enough to show that this map is onto, i.e. every element x of A is of the form $x = \lambda e$ for some $\lambda \in \mathbb{C}$. So take $x \in A$. By the above, $\exists\, \lambda_0 \in \sigma(x)$. Then $x - \lambda_0 e$ is not invertible. But, by assumption, 0 is the only vector in A which is not invertible. Then $x = \lambda_0 e$.

Remark. The argument proving the non-emptiness of the spectrum of $x \in A$, actually shows that $\lambda \mapsto (x-\lambda e)^{-1}$ is an analytic function on the set $\Omega_x := \mathbb{C}\backslash\sigma(x)$. It vanishes at infinity.

Spectral radius formula. For $x \in A$, the spectral radius $\rho(x)$ of x is defined to be the radius of the smallest closed disc centred at $0 \in \mathbb{C}$ which contains $\sigma(x)$. Thus, $\rho(x) = \sup\{|\lambda| \,:\, \lambda \in \sigma(x)\}$. We have the formula

$$\rho(x) = \lim_{n\to\infty} \|x^n\|^{1/n} = \inf_{n\geq 1} \|x^n\|^{1/n}.$$

Proof. We have seen that

$$(\lambda e - x)^{-1} = \sum_{n=0}^{\infty} \lambda^{-n-1} x^n, \quad |\lambda| > \|x\|.$$

Therefore, for any $r > \|x\|$, integrating this formula over the circle Γ_r of radius r, centred at 0, we get $\frac{1}{2\pi i}\int_{\Gamma_r} (\lambda e - x)^{-1}\lambda^n d\lambda = x^n$.

By Cauchy's fundamental theorem, this formula continues to hold for $r > \rho(x)$.

Hence $\|x^n\| \leq \frac{r^n}{2\pi}\int_{\Gamma_r} \|(\lambda e - x)^{-1}\| |d\lambda|$, $r > \rho(x)$.

But $\lambda \mapsto (\lambda e - x)^{-1}$ is bounded on Γ_r, being continuous. So $\exists\, M(r) < \infty \ni \|(\lambda e - x)^{-1}\| \leq M(r)\ \forall\, \lambda \in \Gamma_r$.

Hence $\|x^n\| \leq \frac{M(r) r^n}{2\pi}\int_{\Gamma_r} |d\lambda| = r^n\, M(r)\, r$

So $\|x^n\|^{1/n} \leq r\, (rM(r))^{1/n}$.

Taking limit, $\limsup_{n\to\infty} \|x^n\|^{1/n} \leq r$.

Since this holds for all $r > \rho(x)$, we get $\limsup_{n\to\infty} \|x^n\|^{1/n} \leq \rho(x)$.

On the other hand, for $n = 1, 2, \ldots$ the formula

$$x^n - \lambda^n e = (x - \lambda e)(x^{n-1} + \lambda x^{n-2} + \ldots + \lambda^{n-1} e)$$

$$\text{shows that } \lambda \in \sigma(x) \Rightarrow \lambda^n \in \sigma(x^n) \quad \text{(why?)}$$

$$\Rightarrow |\lambda|^n = |\lambda^n| \leq \|x^n\|$$

$$\Rightarrow |\lambda| \leq \|x^n\|^{1/n}.$$

Thus $\rho(x) \leq \inf_{n\geq 1} \|x^n\|^{1/n}$.

Combining the two inequalities, we get the spectral radius formula.

Exercise. Show that the map $x \mapsto \sigma(x)$ from the Banach algebra A into the collection of non-empty compact subsets of $\mathbb{C}$ is continuous in the following sense. For any open set $\mathbf{U} \supseteq \sigma(x)$, there is a neighbourhood $V \subseteq A$ of x such that $y \in V \Rightarrow \sigma(y) \subseteq \mathbf{U}$.

3. Riesz functional calculus

If A is a Banach algebra, $x \in A$ and $f(z) = \sum_{n=0}^{K} a_n z^n$ is a polynomial over $\mathbb{C}$, then $f(x)$ makes obvious sense: $f(x) := \sum_{n=0}^{K} a_n x^n \in A$. More generally, if $f = \frac{P}{Q}$

is a rational function over $\mathbb{C}$, and f has no pole in $\sigma(x)$ (i.e. Q is non-vanishing on $\sigma(x)$) then one easily sees that $Q(x)$ is invertible, hence one naturally defines $f(x)$ as $P(x)Q(x)^{-1}$. We now wish to show that -more generally- one can define $f(x)$ for any complex valued function which is analytic on some neighbourhood of $\sigma(x)$. However, to avoid confusion, we shall call it $\tilde{f}(x)$ rather than $f(x)$. First we need a couple of technical lemmas.

Lemma. *Let $x \in A$ and $m \in \mathbf{Z}$. Let γ be a simple closed curve in $\mathbb{C}$ which encloses $\sigma(x)$ in the positive sense. Then we have $\frac{1}{2\pi i}\int_\gamma \lambda^m(\lambda e - x)^{-1}d\lambda = x^m$ for $m = 0, 1, 2, \ldots$ If, further, x is invertible then the same formula holds for $m = -1, -2, \ldots$ as well (provided γ does not enclose 0).*

Proof. We first observe that if Q is a compact Hausdorff space and $g : Q \to A$ is a continuous function, then for any $x \in A$, and any Borel complex measure μ on Q, we have

$$x \cdot \int_Q g d\mu = \int_Q (x \cdot g) d\mu, \left(\int_Q g d\mu\right) \cdot x = \int_Q g \cdot x d\mu \qquad (*)$$

Here, naturally, $x \cdot g$ denotes the function $\lambda \mapsto x\, g(\lambda)$, $\lambda \in Q$, and similarly for $g \cdot x$.

We prove the first equality in $(*)$. The second is proved similarly. Consider the bounded linear map $L_x : A \to A$ given by $L_x(y) = xy,\ y \in A$. Then we need to show

$$L_x\left(\int_Q g\, d\mu\right) = \int_Q L_x \circ g\, d\mu.$$

In view of the definition of the integral $\int_Q L_x \circ g\, d\mu$, this will follow once we show that

$$\varphi\left(L_x\left(\int_Q g\, d\mu\right)\right) = \int_Q \varphi \circ L_x \circ g\, d\mu$$

for any $\varphi \in A^*$. That is,

$$(\varphi \circ L_x)\left(\int_Q g\, d\mu\right) = \int_Q \varphi \circ L_x \circ g\, d\mu.$$

But, the definition of the integral $\int_Q g\, d\mu$ implies that this is true since $\varphi \circ L_x \in A^*$. This proves the claim $(*)$.

Now, for $m = 0, 1, 2, \ldots$, (and also for $m = -1, -2, \ldots$ in case x is invertible) let's define the vector $v_m \in A$ by $v_m = \frac{1}{2\pi i}\int_\gamma \lambda^m(\lambda e - x)^{-1}d\lambda$.

In view of Cauchy's fundamental theorem (for vector-valued analytic functions), we may assume that the simple closed curve γ enclosing $\sigma(x)$ lies in the domain $\{\lambda : |\lambda| > \|x\|\}$. But, for λ in this domain we have the expansion

$$\lambda^m(\lambda e - x)^{-1} = \sum_{n=0}^{\infty} \lambda^{m-n-1}x^n$$

which converges uniformly on compact subset of this domain. Therefore, we have, for $m = 0, 1, 2, \ldots$

$$v_m = \sum_{n=0}^{\infty} \frac{1}{2\pi i} \int_\gamma \lambda^{m-n-1} x^n d\lambda = \sum_{n=0}^{\infty} \left(\frac{1}{2\pi i} \int_\gamma \lambda^{m-n-1} d\lambda \right) x^n.$$

Here, in the last step we have used the claim $(*)$. But

$$\frac{1}{2\pi i} \int_\gamma \lambda^{m-n-1} d\lambda = \begin{cases} 1 & \text{if } n = m \\ 0 & \text{if } n \neq m. \end{cases}$$

Therefore $v_m = x^m$ for $m \geq 0$.

Now suppose x is invertible. Then we have $e = -x^{-1}(\lambda e - x) + \lambda x^{-1}$. Hence, for $\lambda \notin \sigma(x)$, $(\lambda e - x)^{-1} = -x^{-1} + \lambda x^{-1}(\lambda e - x)^{-1}$. Hence $\lambda^m(\lambda e - x)^{-1} = -\lambda^m x^{-1} + \lambda^{m+1} x^{-1}(\lambda e - x)^{-1}$. Contour integrating this trivial identity over γ, we get:

$$\begin{aligned} v_m &= -\frac{1}{2\pi i} \int_\gamma \lambda^m x^{-1} d\lambda + \frac{1}{2\pi i} \int_\gamma x^{-1} \cdot \lambda^{m+1} (\lambda e - x)^{-1} d\lambda \\ &= \left(-\frac{1}{2\pi i} \int_\gamma \lambda^m d\lambda \right) x^{-1} + x^{-1} \left(\frac{1}{2\pi i} \int_\gamma \lambda^{m+1} (\lambda e - x)^{-1} d\lambda \right). \end{aligned}$$

Here we have once again used $(*)$.

But, as γ does not enclose 0, $\int_\gamma \lambda^m d\lambda = 0$. Also, the second integral above is v_{m+1}. Thus we get the recurrence relation $v_{m+1} = x v_m$. Since we already know $v_0 = e$, solving this recurrence relation we obtain $v_m = x^m$ for all $m \in \mathbf{Z}$.

Lemma. *Let A be a Banach algebra, $x \in A$, and f be a rational function over $\mathbb{C}$ which has no poles in $\sigma(x)$. Then, for any simple closed curve γ which encloses $\sigma(x)$ in the positive sense but does not enclose any pole of f, we have the formula*

$$f(x) = \frac{1}{2\pi i} \int_\gamma f(\lambda)(\lambda e - x)^{-1} d\lambda.$$

Proof. Fix $x \in A$. Let V denote the $\mathbb{C}$-vector space of all rational functions over $\mathbb{C}$ which have no poles in $\sigma(x)$. Let $V_0 \subseteq V$ consist of all rational functions f satisfying the above formula. Clearly V_0 is a vector subspace of V. Also, from school algebra we know that the set $S = \{z \mapsto (z - \alpha)^m : \alpha \in \mathbb{C} \backslash \sigma(x), m \in \mathbf{Z}\}$ spans V (partial fraction expansion). So, to conclude $V_0 = V$, enough to show that $S \subseteq V_0$. That is, we need to show that for any $\alpha \in \mathbb{C} \backslash \sigma(x)$ and $m \in \mathbf{Z}$, the function $z \mapsto (z - \alpha)^m$ is in V_0. This follows from the previous lemma applied to the vector $x - \alpha e$.

For any domain $\Omega \subseteq \mathbb{C}$, let A_Ω denote the set of vectors $x \in A$ for which $\sigma(x) \subseteq \Omega$. By continuity of spectrum, A_Ω is an open subset of A. Also, Ω is included in A_Ω via the identification map $\lambda \mapsto \lambda e$. Let $\mathrm{H}(\Omega)$ denote the topological vector space of all complex-valued analytic functions on Ω, provided with the topology of locally uniform convergence. (A sequence $\{f_n\}$ in $H(\Omega)$ is said to be locally uniformly converging to $f \in H(\Omega)$ if each $\lambda_0 \in \Omega$ has a neighbourhood $\mathbf{U} \subseteq \Omega$ such that $f_n(\lambda) \to f(\lambda)$ uniformly for $\lambda \in \mathbf{U}$.) In terms of these notations, the previous lemma may be restated as saying that

Lemma. *If $f \in H(\Omega)$ is given by a rational function, then for any $x \in A_\Omega$ and any simple closed curve γ in Ω which positively surrounds $\sigma(x)$, we have*

$$f(x) = \frac{1}{2\pi i} \int_\gamma f(\lambda)(\lambda e - x)^{-1} d\lambda.$$

This prompts us to:

Definition. *For* any $f \in H(\Omega)$ *and* $x \in A_\Omega$ *define* $\tilde{f}(x) \in A$ *by the formula* $\tilde{f}(x) = \frac{1}{2\pi i} \int_\gamma f(\lambda)(\lambda e - x)^{-1} d\lambda$. *Here* γ *is any simple closed curve in* Ω *which encloses* $\sigma(x)$ *in the positive sense.*

Notice that, in view of Cauchy's fundamental theorem, this integral does not depend on the choice of γ.

For any $f \in H(\Omega)$, *this defines a function* $\tilde{f} : A_\Omega \to A$. *Let* $\tilde{H}(A_\Omega) := \{\tilde{f} : f \in H(\Omega)\}$. *Equip* $\tilde{H}(A_\Omega)$ *with the topology of locally uniform convergence. (Warning: If* A *is infinite dimensional then* A_Ω *is* not *locally compact, so this topology is not the topology of uniform convergence on compacts!)*

Now, the map $H(\Omega) \overset{\sim}{\to} \tilde{H}(A_\Omega)$ *satisfies:*

1) $\sim$ *is continuous (Exercise),*
2) $\sim$ *is a linear map (immediate from the linearity of integral)*
3) $\tilde{f}(x) = f(x)$ *for rational functions* $f \in H(\Omega)$, $x \in A_\Omega$. *(previous lemma)*
4) Indeed, $\sim$ *is an algebra homomorphism, (see below), and*
5) If we identify Ω *with the subset* $\{\lambda e : \lambda \in \Omega\}$ *of* A_Ω *(via the map* $\lambda \mapsto \lambda e$*) then* $\tilde{f} \in \tilde{H}(A_\Omega)$ *is an extension of* f. *(Easy exercise).*

To see (4), note that in view of (3), we have $\widetilde{f_1 f_2} = \tilde{f_1}\tilde{f_2}$ *for rational functions* $f_1, f_2 \in H(\Omega)$. *Since by Runge's theorem such rational functions are dense in* $H(\Omega)$, *(1) now implies that* $\widetilde{f_1 f_2} = \tilde{f_1}\tilde{f_2}$ *for all* $f_1, f_2 \in H(\Omega)$.

The Riesz functional calculus (which interprets $f(x)$ as a vector in the Banach algebra for any vector x and complex analytic functions f on $\sigma(x)$) may be used to deduce algebraic information about vectors from information about their spectra. For instance, we have

Theorem. *Let* $x \in A$ *be such that* x *is invertible and* $\sigma(x)$ *does not separate* 0 *and* ∞. *Then* $x = e^y$ *for some* $y \in A$. *In consequence* x *has roots of all orders.*

Proof. The hypothesis implies that there is a planar domain $\Omega \supset \sigma(x)$ such that $0 \notin \Omega$ and Ω is simply connected (why?). Hence there is an analytic branch of the logarithm on Ω, i.e. an $f \in H(\Omega) \ni e^{f(\lambda)} = \lambda$ for all $\lambda \in \Omega$. Now put $y = \tilde{f}(x)$. It easily follows that $x = e^y$.

4. Commutative Banach algebras

Gelfand Theory. *Let* A *be a commutative Banach algebra. By an* ideal *in* A *we shall always mean a proper ideal. Then* $I \subset A$ *is an ideal iff* I *is a linear subspace of* A *such that for* $x \in I$ *and* $y \in A$, $xy \in I$. *Because of commutativity of* A, *ideals are*

two-sided. Note that $e \notin I$ and in fact I is contained in the proper closed subset $A \backslash G$ of A, where G is the group of invertible elements of A. In consequence, whenever I is an ideal, its closure $\bar{I}$ is also a proper ideal. In particular, all maximal ideals are closed. Also note that for each $x \in A \backslash G$, $I = xA$ is a proper ideal containing x (it is proper since $e \notin I$) and hence by a Zorn's lemma argument, any $x \in A \backslash G$ is contained in a maximal ideal.

Thus maximal ideals are plentiful in a commutative Banach algebra A: their union is the set $A \backslash G$ of non-invertible elements. (Recall that, by the Gelfand-Mazur theorem, $\mathbb{C}$ is the only commutative Banach algebra which has no non-invertible element except 0.)

By a *complex homomorphism* of the commutative Banach algebra A, we shall mean an algebra homomorphism $\varphi : A \to \mathbb{C}$, such that $\varphi \not\equiv 0$. $\Delta = \Delta_A$ will denote the set of all complex homomorphisms of A. Notice that if φ is a complex homomorphism then its kernel is a maximal ideal, hence closed. Since a linear functional with a closed kernel is necessarily continuous, it follows that complex homomorphisms are automatically bounded linear functionals. Also, if m is any maximal ideal in A, then A/m is a Banach algebra which is also a field, whence by Banach-Mazur, there is a Banach algebraic isometric isomorphism $A/m \cong \mathbb{C}$. Composing this isomorphism with the canonical homomorphism $A \to A/m$, we get a $\varphi \in \Delta_A$ such that $m = \ker(\varphi)$. Thus $\varphi \mapsto \ker \varphi$ provides a natural bijection between the set Δ_A of complex homomorphism on A and the set of all maximal ideals of A. For this reason, $\Delta = \Delta_A$ is often called the *maximal ideal space* of A.

Lemma. *For $x \in A$, $\sigma(x) = \{\varphi(x) : \varphi \in \Delta\}$. In consequence, we have $|\varphi(x)| \leq \rho(x) \leq \|x\|$ for $\varphi \in \Delta$. Since $\varphi(e) = 1$ and $\|e\| = 1$ for $\varphi \in \Delta$, it follows that $\|\varphi\| = 1$ for all $\varphi \in \Delta$.*

Proof. Recall that $\rho(x)$ denotes the spectral radius of x. Thus only the first statement needs a proof. From the above discussion, $x \in A$ is non-invertible iff x is in a maximal ideal iff $\varphi(x) = 0$ for some $\varphi \in \Delta$. Applying this observation to $x - \lambda e$ when $x \in A$, $\lambda \in \mathbb{C}$, we find that $\lambda \in \sigma(x) \Leftrightarrow \varphi(x) - \lambda = \varphi(x - \lambda e) = 0$ for some $\varphi \in \Delta \Leftrightarrow \varphi(x) = \lambda$ for some $\varphi \in \Delta$.

By the lemma, Δ is a subset of the closed unit ball of the Banach space dual A^* of A. Recall that the weak* topology on A^* is simply the topology of pointwise convergence (A^* after all is a family of functions on A), and Banach Alaoglu theorem says that the unit ball of A^* is compact in this topology. (The unit ball of A^* may be viewed as the subset B of $\hat{\mathbb{C}}^A$ - the space of all functions from A into the Riemann sphere $\hat{\mathbb{C}}$ - given by $B = \{\varphi \in \hat{\mathbb{C}}^A : \varphi(\lambda x + \mu y) = \lambda\varphi(x) + \mu\varphi(y)\ \forall\, x, y \in A,\ \lambda, \mu \in \mathbb{C}$, and $|\varphi(x)| \leq \|x\|$ for all $x \in A\}$. Being a closed subset of the compact space $\hat{\mathbb{C}}^A$, the unit ball is weak*-compact.) Since Δ is clearly a closed subset (in the weak* topology) of the unit ball of A^*, it follows that the subspace (or inherited) topology on Δ is compact Hausdorff. This is called the *Gelfand topology* on Δ. We repeat: Δ with the Gelfand topology becomes a compact Hausdorff space.

For $x \in A$, let $\hat{x} : \Delta \to \mathbb{C}$ denote the function defined by $\hat{x}(\varphi) = \varphi(x)$, $\varphi \in \Delta$. (This is just the restriction to Δ of the canonical isometric copy $\hat{x}$ of x in A^{**}.) From the definition of the Gelfand topology on Δ, it is immediate that $\hat{x}$ is continuous on Δ, i.e. $\hat{x} \in C(\Delta)$. This defines a map $A \overset{\wedge}{\to} C(\Delta)$, called the *Gelfand Transform*. Since each $\varphi \in \Delta$ is an algebra homomorphism, it is immediate that $\wedge$ is an algebra homomorphism from the commutative Banach algebra A into the commutative Banach algebra $C(\Delta)$ (with the supremum norm $\|\cdot\|_\infty$). The last lemma may be re-stated as :

$$\sigma(x) = \text{range}(\hat{x}), \ \|\hat{x}\|_\infty = \rho(x) \leq \|x\|, \ \ x \in A.$$

Thus, $\wedge$ is a continuous algebra homomorphism from A into $C(\Delta)$. But, generally, it is neither 1-1 nor onto. The range of $\wedge$ need not even be closed in $C(\Delta)$.

The *radical* $\text{rad}(A)$ of A is defined as the intersection of all the maximal ideals in A. One says that A is *semi-simple* if $\text{rad}(A) = \{0\}$. Clearly $\text{rad}(A) = \ker(\wedge)$. Thus $\wedge$ is 1-1 iff A is semi-simple.

Example 1. Let K be a compact Hausdorff space. For each $x \in K$, the map $\varphi_x : C(K) \to \mathbb{C}$ given by $\varphi_x(f) = f(x)$, $f \in C(K)$, is a complex homomorphism of $C(K)$. Conversely, if $\varphi \in \Delta$, let $I = \ker(\varphi)$. For $f \in I$, let $\mathbf{Z}(f)$ be the set of zeros of f. If for $f_1, \ldots, f_n \in I$, $\mathbf{Z}(f_1) \cap \ldots \mathbf{Z}(f_n) = \phi$ then $f := \sum_{i=1}^{x} \bar{f}_i f_i \in I$ is invertible in $C(K)$, a contradiction. So the family $\{\mathbf{Z}(f) : f \in I\}$ of compact subsets of K have the finite intersection property. So $\exists\, x \in K \ni x \in \mathbf{Z}(f)\, \forall\, f \in I$. So $I \subseteq \ker(\varphi_x)$ and hence $\ker(\varphi) = I = \ker(\varphi_x)$. So $\varphi = \varphi_x$. Thus $\Delta = \{\varphi_x : x \in K\}$ and the map $x \mapsto \varphi_x$ is a homeomorphism between K and Δ. If one identifies Δ with K via this map, then the Gelfand transform in this case is just the identity map on $C(K)$.

Example 2. Let A denote the set of all functions $f : \mathbb{R} \to \mathbb{C}$ given by an absolutely convergent Fourier series $f(x) = \sum_{n=-\infty}^{\infty} a_n e^{2\pi i n x}$. It is an algebra with point-wise operations. Norm it by : $\|f\| = \sum_{n=-\infty}^{\infty} |a_n|$. Then A becomes a commutative Banach algebra. Note that the subalgebra generated by the element $x \mapsto e^{2\pi i x}$ is dense in A, so that any complex homomorphism on A is determined by its value at this particular element. In consequence, it is easy to see that, as in Example 1, $\Delta = \{\varphi_x : x \in \mathbb{R}\}$ when φ_x is evaluation at $x \in \mathbb{R} : \varphi_x(f) = f(x)$. However, now, all $f \in A$ are periodic with period 1, so that $\varphi_{x+n} = \varphi_x\ \forall\, n \in \mathbf{Z}$. Thus the map $x \mapsto \varphi_x$ from $\mathbb{R}$ to Δ descends to a bijection from $\mathbb{R}/\mathbf{Z}$ to Δ. But $\mathbb{R}/\mathbf{Z}$ is just the unit circle S^1 in $\mathbb{C}$, and it is easy to see that we have got a homeomorphism between S^1 and Δ. The image $\hat{A}$ of A under the Gelfand transform consists of all continuous functions on S^1 which are given by absolutely convergent Fourier series. In particular, it contains all trigonometric polynomials on S^1. Thus, in this case the Gelfand transform is 1-1 and has dense range in $C(\Delta)$, but is not onto.

(Example 2 above may be used to give a quick proof of a famous result from Fourier Analysis:

Wiener's Lemma. *Let $f : \mathbb{R} \to \mathbb{C}$ be given by an absolutely convergent Fourier Series. Suppose $f(x) \neq 0$ for all $x \in \mathbb{R}$. Then $\frac{1}{f} : \mathbb{R} \to \mathbb{C}$ is also given by an absolutely convergent Fourier Series.*

Proof. Let A be the Banach Algebra of Example 2. In view of Example 2, the hypothesis means that $f \in A$ satisfies $\varphi(f) \neq 0\ \forall\, \varphi \in \Delta$, while the conclusion means that f is invertible in A.)

In view of such pathologies in the behaviour of the Gelfand transform in general, we look for special classes of commutative Banach algebras in which this transform is well-behaved. The class of (commutative) C^*-*algebras* is one such class.

Definition. *Let A be a (not necessarily commutative) Banach algebra. A map* $A \xrightarrow{*} A$ *is called an* involution *on A if it is conjugate linear, anti-multiplicative and of order 2. That is,* $(\lambda x + \mu y)^* = \bar{\lambda} x^* + \bar{\mu} y^*$, $(xy)^* = y^* x^*$ *and* $x^{**} = x$ *for all* x, y *in* A, λ, μ *in* $\mathbb{C}$.

Definition. *A* C^**-algebra is a Banach algebra with an involution* $*$ *which satisfies the so-called* C^**-condition :* $\|xx^*\| = \|x\|^2$ *for all vectors* x.

Remark. If $*$ is an involution on a commutative semi-simple Banach algebra then $*$ is automatically continuous. (Hint: By a modification of the closed graph theorem, suffices to show that $x_n \to x$, $x_n^* \to y$ implies $y = x^*$. Now take any $\varphi \in \Delta$, define $\tilde{\varphi}$ by $\tilde{\varphi}(x) = \overline{\varphi(x^*)}$. Then $\tilde{\varphi} \in \Delta$. Thus both φ and $\tilde{\varphi}$ are continuous. Conclude that $\varphi(y) = \varphi(x^*)$. Since this holds for all $\varphi \in \Delta$, semi-simplicity implies $y = x^*$.)

Example 1. For any compact Hausdorff space K, $C(K)$, with conjugation as involution, is a commutative C^*-algebra.

Example 2. For any Hilbert space $\mathcal{H}$, the space $\mathcal{L}(\mathcal{H})$ of bounded operators from $\mathcal{H}$ to $\mathcal{H}$, with the usual adjoint operation as involution, is a C^*-algebra which is non-commutative when $\dim(\mathcal{H}) > 1$.

If A is a Banach algebra with involution $*$, then we say that $x \in A$ is *hermitian* if $x = x^*$. Note that, by uniqueness of the multiplicative identity, e is hermitian (since one easily sees that e^* also is a multiplicative identity). Also, for any $x \in A$, xx^* and $x + x^*$ are hermitian. In Example 1, the hermitian elements in $C(K)$ are just the real-valued continuous functions on K, while in Example 2, an element of $\mathcal{L}(\mathcal{H})$ is hermitian if it is a self-adjoint operator on $\mathcal{H}$. Note that each $x \in A$ can be written uniquely as $x = u + iv$, with u, v hermitian (viz. $u = \frac{x+x^*}{2}$, $v = \frac{x-x^*}{2i}$).

The following important theorem says that the only commutative C^*-algebras are the ones in Example 1.

Gelfand-Naimark Theorem. *If A is a commutative* C^**-algebra then the Gelfand transform is a* $*$*-preserving isometric isomorphism from A onto* $C(\Delta)$.

Proof. The adjective $*$-preserving means that $\wedge$ carries the involution $*$ on A to the involution (conjugation) on $C(\Delta)$. That is, $\widehat{x^*} = \bar{\hat{x}}$ for all $x \in A$. Let's prove this first. To begin with, let's assume x is hermitian: $x^* = x$. Then we need to show that $\hat{x}$ is real-valued, i.e. $\varphi(x) = \hat{x}(\varphi) \in \mathbb{R}$ for all complex homomorphisms φ on A. So take $\varphi \in \Delta$, $x = x^*$ and let $\varphi(x) = a + bi$, $a, b \in \mathbb{R}$. So we must show $b = 0$. Now for any real number t, $\varphi(x + ite) = \varphi(x) + it = a + i(b + t)$. Thus,

$$
\begin{aligned}
|\varphi(x)|^2 + 2bt + t^2 = a^2 + (b+t)^2 = |\varphi(x + ite)|^2 &\leq \|x + ite\|^2 \\
&= \|(x + ite)(x + ite)^*\| = \|xx^* + t^2 e\| \leq \|xx^*\| + \|t^2 e\| \\
&= \|x\|^2 + t^2.
\end{aligned}
$$

Hence, $2bt \leq \|x\|^2 - |\varphi(x)|^2$ for all $t \in \mathbb{R}$. Since the right hand side is independent of t, this forces $b = 0$. Thus $\varphi(x) = a \in \mathbb{R}$ whenever x is hermitian. Next take an arbitrary vector $x \in A$. Write $x = u + iv$ with u, v hermitian. Then $x^* = u - iv$. Hence, for $\varphi \in \Delta$,

$$\begin{aligned}\widehat{x^*}(\varphi) &= \varphi(x^*) = \varphi(u - iv) = \varphi(u) - i\varphi(v) \\ &= \overline{\varphi(u) + i\varphi(v)} \quad (\text{ as } \varphi(u), \varphi(v) \in \mathbb{R}) \\ &= \overline{\varphi(x)} = \overline{\hat{x}(\varphi)}.\end{aligned}$$

Then $\hat{x^*} = \bar{\hat{x}}$ for all $x \in A$.

Now we show that $\wedge$ is an isometry: $\|x\| = \|\hat{x}\|_\infty = \rho(x)$. Again we first show this for hermitian x. If x is hermitian, then so are all powers of x. In particular x^{2^k} is hermitian for $k = 0, 1, 2 \ldots.$ Therefore

$$\|x^{2^{k+1}}\| = \|x^{2^k} \cdot x^{2^k}\| = \|x^{2^k}\|^2.$$

Hence we get $\|x^{2^k}\| = \|x\|^{2^k}$ or $\|x^{2^k}\|^{1/2^k} = \|x\|$ for all k. But the sequence $\{\|x^{2^k}\|^{1/2^k} = \|x\|\}$ is a subsequence of the sequence $\{\|x^n\|^{1/n}\}$, and – by the spectral radius formula – the latter sequence converges to $\rho(x) = \|\hat{x}\|_\infty$. Thus $\|x\| = \|\hat{x}\|_\infty$ for hermitian x. Now, for an arbitrary $x \in A$, xx^* is hermitian, so that

$$\begin{aligned}\|x\|^2 &= \|xx^*\| = \|\widehat{xx^*}\|_\infty = \|\hat{x} \cdot \bar{\hat{x}}\|_\infty \\ &= \|\,|\hat{x}|^2\|_\infty = \|\hat{x}\|_\infty^2.\end{aligned}$$

Thus the Gelfand transform is an isometry. We have already seen that it is an algebra homomorphism. It remains to show that it maps A onto $C(\Delta)$: $\hat{A} = C(\Delta)$. Now, $\hat{A}$ is a subalgebra of $C(\Delta)$. Since $\wedge$ is a $*$-isomorphism, $\hat{A}$ is closed under conjugation. Since $\hat{e}$ is the constant function 1, $\hat{A}$ contains constants. Also, as $\wedge$ is an isometry, it is 1-1. Thus A is semi-simple, i.e., $\hat{A}$ separates points of Δ. Therefore by Stone-Weirstrass theorem $\hat{A} = C(\Delta)$. This completes the proof.

Example. Let m be Lebesue measure on $[0, 1]$, and consider the Banach algebra $L^\infty(m)$ of m-essentially bounded complex-valued measurable functions on $[0, 1]$ with the usual identifications. This is a commutative C^*-algebra with conjugation as the involution. Therefore, by Gelfand -Naimark, $L^\infty(m)$ is isometrically $*$-isomorphic with $C(\Delta)$ for a compact Hausdorff space Δ, via the Gelfand transform $\wedge$. Being $*$-preserving, $\wedge$ maps positive functions in $L^\infty(m)$ to positive functions in $C(\Delta)$. (Generally, given a C^*-algebra A, one says that $x \in A$ is positive if it can be written as $x = yy^*$. Clearly the positive elements of $L^\infty(m)$ and $C(\Delta)$ are the non-negative real-valued functions in the respective space.) In consequence, the map $\hat{f} \to \int_0^1 f\,dm$ is a positive functional on $C(\Delta)$; clearly it maps the constant function $1 \in C(\Delta)$ to 1. Hence by Riesz representation theorem, there is a regular Borel probability measure $\hat{m}$ on Δ such that $\int_0^1 f\,dm = \int_\Delta \hat{f}\,d\hat{m}$ for all $f \in L^\infty(m)$. Hence we can consider the C^*-algebra $L^\infty(\hat{m})$ as well!

Now take any $f \in L^\infty(\hat{m})$. Say $|f(x)| \le c\ (a.e.\ \hat{m})$. By Lusin's theorem, there is a sequence $\{f_n\}$ of continuous functions on Δ such that $|f_n(x)| \le c$ for all x and $f_n \to f$ in the norm of $L^2(\hat{m})$. Since $\wedge$ is onto, there is a sequence $\{g_n\}$ in $L^\infty(m)$ such that $\hat{g_n} = f_n$ for all n. Since $\wedge$ preserves conjugation, we have $(|g_m - g_n|^2)^\wedge = |f_m - f_n|^2$ and hence, by the defining property of $\hat{m}$, we have $\int_0^1 |g_m - g_n|^2 dm = \int_\Delta |f_m - f_n|^2 d\hat{m}$. Thus $\{g_n\}$ is Cauchy in $L^2(m)$. Also, since $\wedge$ preserves positivity, from $|f_n(x)| \le c\ \forall\ x \in \Delta$, we can conclude that $|g_n(x)| \le c\ \forall\ x \in [0, 1]$. Therefore there is a $g \in L^2(m)$ such that $g_n \to g$ in $L^2(m)$, and $|g(x)| \le c\ (a.e.\ m)$. Thus $g \in L^\infty(m)$. Now, from the previous observation that $\|g_m - g_n\|_2 = \|f_m - f_n\|_2 \to 0$, we conclude that $f_m = \hat{g}_m \to \hat{g}$. But, replacing $\{f_m\}$ by a suitable subsequence if necessary, we may assume that $f_m \to f(a.e.\ \hat{m})$. Thus $f = \hat{g} \in C(\Delta)(a.e.\ \hat{m})$. Thus every element $f \in L^\infty(\hat{m})$ is a.e. equal to an element of $C(\Delta) \cdot$ $(\Delta, \hat{m})$ has the curious property that $L^\infty(\hat{m}) = C(\Delta)$.

Exercise. Let A be a commutative C^*-algebra. Suppose there is an $x_0 \in A$ such that the subalgebra of A generated by x_0 and x_0^* is dense in A. Then show that the map $\Delta \to \sigma(x_0)$, given by $\varphi \mapsto \varphi(x_0) \in \sigma(x_0)$, is a homeomorphism between Δ and $\sigma(x_0)$. This induces an isometric $*$-isomorphism between $C(\sigma(x_0))$ and $C(\Delta)$. Composing this with the Gelfand transform, we get an isometric $*$-isomorphism between A and $C(\sigma(x_0))$. For any $f \in C(\sigma(x_0))$, let $\tilde{f}(x_0) \in A$ denote its image under this isomorphism. Show that this extends the definition of $\tilde{f}(x_0)$ (Riesz calculus) previously given only for functions f which are analytic on $\sigma(x_0)$. Thus, under the restrictions given in this exercise, one has a "continuous functional calculus".

Now, if B is a (not necessarily commutative) C^*-algebra, and $x_0 \in B$ is *normal* in the sense that $x_0 x_0^* = x_0^* x_0$, then one can apply the above to the closed sub-algebra A spanned by x_0 and x_0^*. Therefore one can develop a continuous functional calculus for normal vectors. One problem is that the spectrum of x_0 as an element of A may not equal its spectrum as an element of B. But there are ways to get around this difficulty. In particular, all these apply to normal operators on Hilbert spaces.

References

[1] B. Bagchi, On Nyman, Beurling and Baez-Duarte's Hilber space reformulation of Riemann Hypothesis, *Proc. Indian Acad. Sci. (Math. Sci.)* **116** no. 2 (2006) 1–10.

[2] W. Rudin, *Real and complex analysis*, Second edition, McGraw-Hill Series in Higher Mathematics. McGraw-Hill Book Co., New York-Düsseldorf-Johannesburg (1974).

[3] W. Rudin, *Functional analysis*, McGraw-Hill Series in Higher Mathematics. McGraw-Hill Book Co., New York-Düsseldorf-Johannesburg (1973) xiii+397 pp.

Proc. Int. Conf. – Lectures on Operator Theory No. 8, 2009, pp. 21–50.

Some Linear Operators and their Invariant Subsets

Gilles Godefroy

Equipe d'Analyse, Universite Paris 6, 4, Place Jussieu, 75252 Paris, Cedex 05, France
e-mail: gig@ccr.jussieu.fr

These notes outline the contents of a series of lectures delivered by the author during a meeting at the Indian Statistical Institute (Bangalore) in June 2007. Some classical and also more recent results on invariant subspaces are displayed in these notes, most of the time with complete proofs: Lomonosov's theorems on the existence of invariant subspaces under commutation assumptions with compact operators, the smooth variational approach to invariant subspaces, the hypercyclic and frequently hypercyclic operators, and the existence of operators with prescribed orbit on arbitrary separable Banach spaces.

Section 6 of these notes is devoted to a Colloquium lecture (also delivered at the Indian Statistical Institute) honoring the late Professor G. Choquet, who was the author's advisor.

Notation. We denote by X a separable infinite-dimensional Banach space, over $\mathbb{R}$ or $\mathbb{C}$. A linear bounded map $T : X \to X$ is called an *operator*, and $L(X)$ denotes the space of such operators. A *subspace* of X is a linear subspace, which is not assumed to be closed unless otherwise specified. A subspace M is non-trivial if $\{0\} \neq M \neq X$, and T-invariant if $T(M) \subseteq M$.

We denote by $K(X)$ the subspace of $L(X)$ consisting of compact operators. The spectrum $\sigma(T)$ of an operator T on a complex space X is

$$\sigma(T) = \{\lambda \in \mathbb{C};\ (T - \lambda I) \notin L(X)^{-1}\}$$

where $L(X)^{-1}$ denotes the group of invertible elements of $L(X)$. The subset of $\sigma(T)$ consisting of the eigenvalues of T is denoted $\sigma_p(T)$. When $\lambda \in \sigma_p(T)$, we denote

$$E_\lambda(T) = \ker(T - \lambda I)$$

the corresponding eigen space.

Otherwise the notation and terminology are classical, or will be defined before use.

1. Lomonosov's theorems

The invariant subspace problem consists into finding, for an operator $T \in L(X)$, a closed non-trivial subspace M such that $T(M) \subseteq M$. In general, this cannot be done

[16], and there even exists an operator on $l_1(\mathbb{N})$ with no non-trivial closed invariant subset [38]. However, it is not known if every operator on a reflexive space X, or on a Hilbert space H, has a closed non-trivial invariant subspace. We refer to [17] for a recent survey of this topic.

A major positive result has been shown by Lomonosov in 1973 [30]. Under a weaker (but already striking) form, this result reads as follows.

Theorem 1.1. *Let X be an infinite-dimensional complex Banach space. Let $T \in K(X)\backslash\{0\}$ be a non-zero compact operator. Then there exists a non-trivial closed subspace M of X such that*

$$S(M) \subseteq M$$

for all $S \in L(X)$ such that $ST = TS$.

In other words, the algebra

$$\Gamma = \{S \in L(X);\ ST = TS\}$$

is *not* transitive. We recall that a subalgebra A of $L(X)$ is said to be transitive when for all non-zero $x \in X$, the space $\{T(x);\ T \in A\}$ is dense in X.

Proof. The original proof by Lomonosov relied on Schauder's fixed point theorem. This simplified version due to T. M. Hilden merely uses the spectral radius formula.

For all $x \in X$, we denote

$$\Gamma(x) = \{S(x);\ ST = TS\};$$

if there exists $x \neq 0$ such that $\overline{\Gamma(x)} \neq X$, then $M = \overline{\Gamma(x)}$ clearly works.

Assume now that $\overline{\Gamma(x)} = X$ for all $x \neq 0$. Pick $x_0 \in X$ such that $Tx_0 \neq 0$, and a ball $B = B(x_0 : r)$ such that for all $x \in B$

$$\left.\begin{aligned} 2\|T(x)\| &\geq \|T(x_0)\| \\ 2\|x\| &\geq \|x_0\| \end{aligned}\right\}$$

The set $\overline{T(B)} = K$ is compact and $0 \notin K$. For all $y \in K$, the set $\Gamma(y)$ is dense, and thus there exists $W \ni y$ open and $S \in \Gamma$ such that $S(W) \subseteq B$. By compactness, K is included into the union of finitely many such open sets, which we write

$$K \subseteq \bigcup_{i=1}^{n} W_i$$

and if $S_i \in \Gamma$ is such that $S_i(W_i) \subseteq B$, we denote

$$\mu = \sup\{\|S_1\|,\ \|S_2\|,\ \ldots,\ \|S_n\|\}$$

Starting from $x_0 \in B$, we have $T(x_0) \in K$, then we find i_1 such that $S_{i_1}T(x_0) \in B$, then again $TS_{i_1}T(x_0) \in K$ and there is i_2 with $S_{i_2}TS_{i_1}T(x_0) \in B$, and so on. If

$$\begin{aligned} x_N &= S_{i_N}TS_{i_{N-1}}T \ldots S_{i_1}T(x_0) \\ &= S_{i_N}S_{i_{N-1}} \ldots S_{i_1}T^N(x_0) \end{aligned}$$

one has

$$\frac{1}{2}\|x_0\| \leq \|x_N\| \leq \mu^n \|T^N\| \, \|x_0\|$$

and hence

$$\lim_{N\to\infty} \|T^N\|^{\frac{1}{N}} = \rho(T) \geq \frac{1}{\mu} > 0.$$

It follows that $\sigma(T) \neq \{0\}$. But since T is compact, $\sigma(T)\backslash\{0\} \subseteq \sigma_p(T)$ and thus T has an eigenvalue $\lambda \neq 0$. By compactness again, $\dim(E_\lambda(T)) < \infty$ and $S(E_\lambda(T)) \subseteq E_\lambda(T)$ if $ST = TS$. Therefore $M = E_\lambda(T)$ works. □

Remarks.

1. Lomonosov's theorem is essentially valid for real Banach spaces. We refer to [26] and [44] for this topic.
2. There exist bounded operators which commute with no $T \in K(X)\backslash\{0\}$. If $R \in L(l^2(\mathbb{Z}))$ is the forward shift defined by

   $$R((x_n)_n) = (x_{n-1})_{n\in\mathbb{Z}}$$

   we have for all $y \in l^2(\mathbb{Z})$

   $$\lim_{n\to\infty} \|R^{2n}(y) - R^n(y)\| = \sqrt{2}\|y\|$$

   if $TR = RT$ and $T(x) = y \neq 0$, it follows

   $$\lim_{n\to+\infty} \|T(R^{2n}(x)) - T(R^n(x))\| = \sqrt{2}\|y\| \tag{1}$$

   but a compact operator maps weakly-null sequences to norm-convergent sequences, and thus if T is compact,

   $$\lim_{n\to+\infty} \|TR^n(x)\| = 0$$

   and this contradicts (1).
3. Theorem 1.1 above is not optimal, since Lomonosov showed that if $ST = TS$ with $T \in K(X)\backslash\{0\}$ and $S \neq \lambda I$, there is a closed non-trivial subspace M such that $R(M) \subseteq M$ for all R such that $RS = SR$. A more general result will be shown below (Corollary 1.4) for reflexive spaces X.
4. By the above, if $GS = SG$ and $ST = TS$, with $T \in K(X)\backslash\{0\}$ and $S \neq \lambda I$, then G has a closed non-trivial invariant subspace. This result does not extend to a chain of 4 operators: it is shown in [45] that if L is Read's operator on l_1 without invariant subspace, there is $T \in K(l_1)\backslash\{0\}$ and $A, B \neq \lambda I$ such that $LA = AL$, $AB = BA$ and $BT = TB$.

Before stating another important result due to V. Lomonosov, let us introduce the following notation: if $T \in L(X)$, its essential norm $|||T|||$ is defined by

$$|||T||| = \inf\{\|T - L\| : L \in K(X)\}.$$

The weak operator topology w_{OT} on $L(X)$ is defined by

$$T_\alpha \xrightarrow{w_{OT}} T \Leftrightarrow \langle x^*, \; T_\alpha(x)\rangle \to \langle x^*, T(x)\rangle$$

for all $x \in X$ and all $x^* \in X^*$. We denote by T^* the adjoint of the operator T, such that

$$\langle x^*, \; T(x)\rangle = \langle T^*(x^*), x\rangle$$

We now have [31]:

Theorem 1.2. *Let E be a complex Banach space. Let $A \subseteq L(E)$ be a proper w_{OT}-closed unital subalgebra. Then there exists $e^* \in E^*\backslash\{0\}$ and $e^{**} \in E^{**}\backslash\{0\}$ such that for all $T \in A$,*

$$|\langle e^{**}, \; T^*(e^*)\rangle| \leq |||T|||.$$

Note that this theorem can be seen as an extension of Burnside's classical theorem asserting that a proper subalgebra of $L(\mathbb{C}^n)$ is not transitive (see [34] for a short proof). We first need the following classical lemma:

Lemma 1.3. *Let X be a Banach space. Let $\mathcal{B} \subseteq L(X)$ be a transitive algebra. If there is $T \in \mathcal{B}$ with*

$$0 < \; rank \; (T) < \infty$$

then $\mathcal{B}$ is w_{OT}-dense in $L(X)$.

Proof. Let $V = T(X)$, and

$$\mathcal{B}_V = \{TR; \;\; R \in \mathcal{B}\}$$

$\mathcal{B}_V$ is a subalgebra of $L(V)$, and since $\mathcal{B}(u)$ is dense for every $u \in V\backslash\{0\}$, it follows that $\mathcal{B}_V$ is a transitive subalgebra of $L(V)$. Now Burnside's theorem asserts that $\mathcal{B}_V = L(V)$, and thus there is $S \in \mathcal{B}_V$ such that $\text{rank}(S) = 1$. Then $(ST) \in \mathcal{B}$ and $\text{rank}(ST) = 1$.

Using transitivity of $\mathcal{B}$ again, we find a dense subspace D of X and a w^*-dense subspace D^* of X^* such that

$$\mathcal{B} \supseteq \{x \otimes x^*; x \in D, \;\; x^* \in D^*\}$$

and it follows that $\mathcal{B}$ is w_{OT}-dense in $L(X)$. □

Proof. Theorem 1.2: We follow the lines of the proof from [29]. Define

$$\phi = \{T \in A; \; |||T||| \leq 1/24\}.$$

If there exists $e^* \in E^*\backslash\{0\}$ such that

$$\{T^*(e^*); \; T \in \phi\}$$

is not dense in $(E^*, \|\cdot\|)$, then the Hahn-Banach theorem easily provides the conclusion. Assume by contradiction that it is not so. It implies in particular that the algebra

$$A^{(*)} = \{T^*; \;\; T \in A\}$$

is transitive. We now have:

Claim. There exists $R \in \phi$ such that $E_1(R^*) = \ker(R^* - I)$ satisfies

$$0 < \dim(E_1(R^*)) < \infty.$$

Proof of the Claim. Pick $e_0^* \in E^*$ with $\|e_0^*\| = 2$ and let $D = \{e^* \in E^*; \|e_0^* - e^*\| \leq 1\}$. For all $e^* \in D$, there is $T_{e^*} \in \phi$ such that

$$\|T_{e^*}^*(e^*) - e_0^*\| \leq \frac{1}{4}.$$

Pick $K_{e^*} \in K(E)$ such that

$$\|T_{e^*} - K_{e^*}\| < \frac{1}{23}$$

Since K_{e^*} is compact, the map

$$K_{e^*}^* : (D, w^*) \to (E^*, \ \|\cdot\|)$$

is continuous. Hence there exists V_{e^*} a w^*-open neighborhood of e^* in D such that

$$\|K_{e^*}^*(m^*) - K_{e^*}^*(e^*)\| < \frac{1}{8}$$

for all $m^* \in V_{e^*}$. We now have

$$\begin{aligned}\|T_{e^*}^*(m^*) - T_{e^*}^*(e^*)\| &\leq \|T_{e^*} - K_{e^*}\| \cdot \|m^*\| + \frac{1}{8} + \|K_{e^*} - T_{e^*}\| \cdot \|e^*\| \\ &\leq \frac{3}{23} + \frac{1}{8} + \frac{3}{23} < \frac{1}{2}\end{aligned}$$

and thus

$$\|T_{e^*}^*(m^*) - e_0^*\| < \frac{3}{4}.$$

This shows that $T_{e^*}^*(V_{e^*}) \subseteq D$. Since D is w^*-compact, there exist $\{e_1^*, e_2^*, \ldots, e_n^*\} \subseteq D$ such that

$$D \subseteq \bigcup_{i=1}^{n} V_{e_i^*}.$$

We now use a non-linear argument: there is a partition of unity subordinated to the covering $\{V_{e_i^*}\}$, in other words continuous functions $f_i : (D, w^*) \to [0, 1]$ such that

$$\begin{cases} \text{(i)} \quad \sum_{i=1}^{n} f_i(e^*) = 1 & \text{for all } e^* \in D \\ \text{(ii)} \quad f_i(e^*) = 0 & \text{if } e^* \in D \backslash V_{e_i^*}. \end{cases}$$

Let us define $g : D \to D$ by

$$g(e^*) = \sum_{i=1}^{n} f_i(e^*) T_{e_i^*}^*(e^*).$$

The map g is continuous, and Schauder's fixed point theorem provides $m^* \in D$ such that $g(m^*) = m^*$. We let now

$$R = \sum_{i=1}^{n} f_i(m^*) T_{e_i^*}.$$

We have $R \in \phi$ and $R^*(m^*) = m^*$, hence $0 < \dim(E_1(R^*))$. On the other hand, $|||R^*||| \leq \frac{1}{24} < 1$, and this easily implies that $\dim(E_1(R^*)) < \infty$. Therefore the claim is shown.

We now conclude the proof of the theorem through some spectral theory. Recall that the essential spectral radius of an operator T is given by

$$\rho_{ess}(T) = \lim_{n \to \infty} |||T^n|||^{1/n} \leq |||T|||$$

and that if $\lambda \in \sigma(T)$ is such that $|\lambda| > \rho_{ess}(T)$ then $\lambda \in \sigma_p(T)$, λ is isolated in $\sigma(T)$ and

$$0 < \dim(\ker(T - \lambda I)) < \infty.$$

Let us note in passing that special cases of these classical assertions were already used twice in these notes. Since $|||R^*||| < 1$, the value $1 \in \sigma(R^*)$ is an isolated point of $\sigma(R^*)$. Let $r > 0$ be such that $|\lambda - 1| > r$ for all $\lambda \in \sigma(R^*)\backslash\{1\}$, and let $G = \{z \in \mathbb{C};\ |1 - z| = r\}$. We have

$$(\lambda I - R^*) \in L(E)^{-1}$$

for all $\lambda \in G$. Since moreover the circle G is contained in the unbounded component of the set $\mathbb{C}\backslash\sigma(R^*)$, we even have (see [41], Th. 10.18) that

$$(\lambda I - R^*) \in A^{(*)^{-1}}$$

for all $\lambda \in G$. If we now let

$$P = \frac{1}{2i\pi} \int_G (\lambda I - R^*)^{-1} d\lambda$$

the map P is a projection from E^* onto $E_1(R^*)$ and by the above $P \in A^{(*)}$. □

Applying our lemma to $X = E^*$, $\mathcal{B} = A^{(*)}$ and $T = P$ shows that $A^{(*)}$ is w_{OT}-dense in $L(E^*)$, and thus A is w_{OT}-dense in $L(E)$. This contradicts our assumption that A is a proper w_{OT}-closed subalgebra of $L(E)$.

Remark. It has been shown by Pearcy and Shields [36] that one can replace the condition "$0 < \text{rank}\,(T) < \infty$" in the above Lemma 1.3 by the condition "$T \in K(X)\backslash\{0\}$". The optimal version of Lomonosov's theorem 1.1. follows since if $S \neq \lambda I$, the commutant algebra $\mathcal{B}$ of S defined by

$$\mathcal{B} = \{R \in L(X) :\ RS = SR\}$$

is not w_{OT}-dense. Hence, if $\mathcal{B}$ contains a non-zero compact operator then $\mathcal{B}$ is not transitive. In particular, an operator which commutes with a non-scalar operator which commutes with a non-zero compact operator has a non trivial closed invariant subspace.

The following corollary implies this Pearcy-Shields result in the case of reflexive spaces X.

Corollary 1.4. *If A is a w_{OT}-closed proper subalgebra of $L(E)$ and if $A \cap K(E) \neq \{0\}$, then the algebra $A^{(*)}$ is not transitive.*

Proof. Let $T \in A \cap K(E)$, $T \neq 0$. We have $|||T_2 T T_1||| = 0$ for every $T_1, T_2 \in A$, and thus it follows from Theorem 1.2. that

$$\langle e^{**}, T_1^* T^* T_2^*(e^*)\rangle = 0.$$

If $\{0\} \neq \cap\{\ker(S^*);\ S \in A\} = M_1$, this space M_1 is non-trivial and $A^{(*)}$-invariant. If not, then either

$$\overline{\{S^*(e^*);\, S \in A\}} = M_2 \neq E^*$$

and then this space M_2 is A^* invariant, or $M_1 = \{0\}$ and $M_2 = E^*$ and then

$$M_3 = \overline{\text{span}}\{T_1^* T^* T_2^*(e^*);\ T_1, T_2 \in A\}$$

is a non-trivial $A^{(*)}$-invariant subspace. □

Remark. This corollary falls short from using the full power of Theorem 1.2. Following [31], let us say that a subalgebra $A \subseteq L(X)$ has the Pearcy-Salinas property if there exists a net $(T_\alpha) \subseteq A$ and an operator $T \in L(X)\backslash\{0\}$ such that

$$\begin{cases} w_{OT} - \lim\limits_{\alpha}(T_\alpha^*) &= T^* \\ \lim\limits_{\alpha} |||T_\alpha||| &= 0 \end{cases}$$

An easy modification of the proof of Corollary 1.4. shows that if A is a w_{OT}-closed proper subalgebra of $L(X)$ with the Pearcy-Salinas property, then the algebra $A^{(*)}$ is not transitive.

It looks like the right place for formulating Lomonosov's conjecture:

Open Problem. Let X be a Banach space, and $T \in L(X)$. Does the adjoint $T^* \in L(X^*)$ have a closed non-trivial invariant subspace?

We may of course assume that X^* is separable. Note that a positive answer to this problem would solve the invariant subspace problem for reflexive spaces. We mention along these lines a result of [42] asserting that Read's operator on l_1 is *not* the adjoint of any operator acting on a predual of l_1.

2. Using the exponential map

In this section, X denotes a separable *real* Banach space with $\dim(X) > 1$. We denote by Exp : $L(X) \to L(X)^{-1}$ the usual exponential map.

With this notation, the following holds [4]:

Theorem 2.1. *Let $A \subseteq L(X)$ be a subalgebra. If there exists a closed subset F of X with $\{0\} \neq F \neq X$ and $Exp(T)(F) \subseteq F$ for every $T \in A$, then the algebra A is not transitive.*

Let us underline the fact that F is not assumed to be a linear subspace, and in the applications of this result, it will indeed not be linear. But a linear subspace will be obtained from F through a differentiation argument.

Proof. This proof relies again on a non-linear technique, and more precisely on the following non-linear counterpart of the classical Bishop-Phelps theorem:

Lemma 2.2. *Let F be a non-trivial closed subset of a separable real Banach space X, with $\dim(X) > 1$. Then there exist $w \in F \backslash \{0\}$ and a Lipschitz Gateauz-Smooth function $H : X \to \mathbb{R}$ such that*

$$\begin{cases} H(w) & = \min\{H(x); x \in F\} \\ H'(w) & \neq 0 \end{cases}.$$

We recall that "H Gateaux-Smooth" means that for all $x \in X$, there exists $H'(x) \in X^*$ such that for all $y \in X$.

$$\lim_{t \to 0} t^{-1}[H(x + ty) - H(x)] = \langle H'(x), y \rangle.$$

Lemma 2.2 is an easy consequence of the smooth variational principle [15]. The restriction $\dim(X) > 1$ permits to avoid the case $F = [0, +\infty) \subseteq \mathbb{R}$ for which the lemma clearly fails.

Pick now $T \in A$, and define $\phi_T : \mathbb{R} \to \mathbb{R}$ by

$$\phi_T(s) = H(\text{Exp}(sT)w).$$

Our assumptions allow us to use the chain rule to show

$$\phi_T'(s) = \langle H'(\phi_T(s)), \ \text{Exp}(sT)T(w) \rangle$$

Since ϕ_T attains its minimum at $s = 0$, one has

$$\phi_T'(0) = 0 = \langle H'(w), \ T(w) \rangle.$$

If $T(w) = 0$ for all $T \in A$, the space $M = \overline{\text{span}} (\{w\})$ is A-invariant. If not, the space $\overline{\{T(w); T \in A\}}$ is a non-trivial closed A-invariant subspace. Hence A is not transitive. □

We denote by $\mathbb{R}[\xi]$ the algebra of real polynomials. The notation $E_T(x)$ defined below will be used from now on.

Corollary 2.3. *Let $T \in L(X)$ be such that there is $x \in X \backslash \{0\}$ such that*

$$E_T(x) = \{Exp(P(T))(x); \ P \in \mathbb{R}[\xi]\}$$

is not dense in X. Then T has a closed non-trivial invariant subspace.

Proof. Let $\overline{E_T(x)} = F(x)$. We apply Theorem 2.1. to the algebra

$$A = \{P(T);\ P \in \mathbb{R}[\xi]\}.$$

Since A is commutative, we clearly have $\mathrm{Exp}(S)(F(x)) \subseteq F(x)$ for all $S \in A$, and Theorem 2.1. concludes the proof. □

Example. An early result due to Krein and Rutman [28] states that if C is a closed, convex, proper cone *with non-empty interior* in X and $T \in L(X)$ satisfies $T(C) \subseteq C$, then T has a closed non-trivial invariant subspace.

Indeed, if $C \cap (-C) = V \neq \{0\}$, then the closed subspace V works. If not pick any $x \in C\backslash\{0\}$. We claim that for every $P \in \mathbb{R}[\xi]$,

$$\mathrm{Exp}(P(T))x \notin (-C). \tag{1}$$

Indeed, let us write $P = P_1 - P_2$ where P_1 and P_2 have positive coefficients. If (1) fails then

$$\mathrm{Exp}(P_2(T))\mathrm{Exp}(P(T)(x)) \in (-C)$$

since C is convex and the series expansion of $\mathrm{Exp}(P_2(\xi))$ has positive coefficients. But this means

$$\mathrm{Exp}(P_1(T))(x) \in (-C)$$

and since on the other hand $\mathrm{Exp}(P_1(\xi))$ also has positive coefficients, we have

$$\mathrm{Exp}(P_1(T))(x) \in C$$

thus $\mathrm{Exp}(P_1(T))(x) = 0$, and this contradicts $x \neq 0$ since $\mathrm{Exp}(P_1(T)) \in L(X)^{-1}$. We have shown that

$$E_T(x) \cap (-C) \neq \emptyset$$

and since $(-C)$ has non-empty interior, $E_T(x)$ is not dense and Corollary 2.3 concludes the proof.

Our next corollary concerns the algebra of multipliers of a closed convex cone C. If C is a proper convex cone and $T, S \in L(X)$, we say that

$$S \leq_C T$$

if $(T - S)(C) \subseteq C$. An operator T is called a C-multiplier if for some $\lambda \geq 0$,

$$-\lambda I \leq_C T \leq_C \lambda I.$$

We denote by A_C the space of C-multipliers. It is easy to check that A_C is a subalgebra of $L(X)$.

Corollary 2.4. *The algebra A_C is not transitive.*

Proof. Pick any $T \in A_C$. There is $\lambda \geq 0$ such that $(T + \lambda I)(C) \subseteq C$. Since C is closed and convex and the exponential series has positive coefficients, we have

$$\mathrm{Exp}(T + \lambda I)(C) \subseteq C.$$

But $\mathrm{Exp}(T + \lambda I) = e^{\lambda}\mathrm{Exp}(T)$, and since $e^{\lambda} > 0$, it follows that

$$\mathrm{Exp}(T)(C) \subseteq C$$

and Theorem 2.1 concludes the proof. □

Example. Let $w = (w_n)_{n\in\mathbb{Z}}$ such that $w_n > 0$ for all n and

$$0 < \inf_{n\in\mathbb{Z}} \frac{w_{n+1}}{w_n} \leq \sup_{n\in\mathbb{Z}} \frac{w_{n+1}}{w_n} < \infty$$

we denote by $l^2_w(\mathbb{Z})$ the space of *complex* sequences (x_n) such that

$$\sum_{n=-\infty}^{+\infty} w_n |x_n|^2 < \infty$$

normed by

$$\|(x_n)\|_w = \left(\sum_{-\infty}^{+\infty} w_n |x_n|^2\right)^{\frac{1}{2}}.$$

Let $R : l^2_w(\mathbb{Z}) \to l^2_w(\mathbb{Z})$ be the forward shift, such that

$$R((x_n)_{n\in\mathbb{Z}}) = (x_{n-1})_{n\in\mathbb{Z}}.$$

The question (still open in full generality, see [2]) is to find a proper closed subspace M of $l^2_w(\mathbb{Z})$ such that $R(M) = R^{-1}(M) = M$, in other words a subspace invariant under the translations of $\mathbb{Z}$. In the case of *even* weights, such that $w_n = w_{-n}$, this can be done as follows: define formally, if $x = (x_n)_{n\in\mathbb{Z}}$,

$$\hat{x}(\theta) = \sum_{n=-\infty}^{+\infty} x_n e^{in\theta}$$

and let $\tilde{x}_n = \overline{x_{-n}}$. With this notation

$$\bar{\hat{x}}(\theta) = \hat{\tilde{x}}(\theta)$$

and thus $\hat{x}(\theta) \in \mathbb{R}$ for all θ if and only if $x_{-n} = \bar{x_n}$ for all n. We consider now

$$X = \{(x_n) \in l^2_w(\mathbb{Z});\ \ x_{-n} = \bar{x_n}\}.$$

The space X is a real Hilbert space. if

$$E = \frac{R + R^{-1}}{2}$$

we have

$$E(\hat{x})(\theta) = \cos\theta.\hat{x}(\theta)$$

and $E(X) \subseteq X$. Similarly, if

$$F = \frac{R - R^{-1}}{2i}$$

we have

$$\hat{F}(x)(\theta) = \sin\theta \cdot \hat{x}(\theta)$$

and $F(X) \subseteq X$. We now order $L(X)$ through

$$C = \{x \in X; \hat{x} \geq 0\}.$$

Since $-1 \leq \cos\theta \leq 1$, we have

$$-I \leq_C E \leq_C I$$

and similarly for F. Hence E and F belong to the non-transitive algebra A_C, and thus X contains a proper closed subspace $\mathcal{Z}$ such that $E(\mathcal{Z}) \subseteq \mathcal{Z}$ and $F(\mathcal{Z}) \subseteq \mathcal{Z}$. Since $R = E + iF$ and $R^{-1} = E - iF$, the existence of a complex closed subspace $M = \mathcal{Z} + i\mathcal{Z}$ of $l_w^2(\mathbb{Z})$ such that $R(M) = R^{-1}(M) = M$ follows.

We refer to [2] and [3] for more general results along the same lines. Before stating the next corollary, we need a definition.

Definition 2.5. *An operator $T \in L(X)$ has a moment sequence if there exists $x_0 \in X\backslash\{0\}$, $x_0^* \in X^*\backslash\{0\}$ and a positive measure $\mu \geq 0$ on $\mathbb{R}$ such that for all $n \geq 0$,*

$$\langle x_0^*, T^n(x_0)\rangle = \int u^n d\mu(u) \tag{2}$$

With this notation, the following holds:

Corollary 2.6. *Every operator which has a moment sequence has a closed non-trivial invariant subspace.*

Proof. Equation (2) and the spectral radius formula show that μ is supported by $[-\|T\|, \|T\|]$. Hence, uniform convergence shows that for every polynomial $Q \in \mathbb{R}[\xi]$,

$$\langle x_0^*, \text{Exp}(Q(T))(x_0)\rangle = \int e^{Q(u)} d\mu(u)$$

and therefore

$$E_T(x_0) \subseteq \{x;\ x_0^*(x) \geq 0\}$$

and Corollary 2.3. concludes the proof. □

Before applying Corollary 2.6. let us observe that it is immediate that, conversely, an operator which has a proper, closed, invariant subspace has a moment sequence (with $\mu = 0$). We denote now $S(X)$ the set of all $T \in L(X)$ such that there exists a compact subset Σ of $\mathbb{R}$ such that

$$|||P(T)||| \leq \sup\{|P(t)|; t \in \Sigma\}.$$

It is clear that $S(X) + K(X) = S(X)$.

For instance, self-adjoint operators S on l^2 belong to $S(l^2)$, with $\Sigma = \sigma(S)$. We can now show ([4]):

Proposition 2.7. *If $T \in S(X)$, then T^* has a closed non-trivial invariant subspace.*

Proof. A modification, also due to Lomonosov, of Theorem 1.2 ([32]), shows that there exists $x_0^* \in X^*\backslash\{0\}$ and $x_0^{**} \in X^{**}\backslash\{0\}$ such that for every $P \in \mathbb{R}[\xi]$

$$|\langle x_0^{**}, P(T^*)(x_0^*)\rangle| \leq |||P(T^*)|||\langle x_0^{**}, x_0^*\rangle$$

and since $T \in S(X)$, we have

$$|\langle x_0^{**}, P(T^*)(x_0^*)\rangle| \leq \sup_{\Sigma} |P| \langle x_0^{**}, x_0^*\rangle.$$

If $\langle x_0^{**}, x_0^*\rangle = 0$, the conclusion clearly holds. If not, we define the following linear form on $\mathbb{R}[\xi]$:

$$\psi(P) = \frac{\langle x_0^{**}, P(T^*)(x_0^*)\rangle}{\langle x_0^{**}, x_0^*\rangle};$$

since $\psi(1) = 1$ and

$$|\psi(P)| \leq \sup_{\Sigma} |P|$$

there is a probability measure $\mu \geq 0$ with support contained in Σ such that

$$\psi(P) = \int P(u) d\mu(u)$$

for all $P \in \mathbb{R}[\xi]$, and thus T^* has a moment sequence. Corollary 2.6 concludes the proof. □

Example. If $S = S^*$ is a self-adjoint operator on a *real* Hilbert space H and $(S - S^*) \in K(H)$ is a compact operator, then $S \in S(H)$ and thus S has closed non-trivial invariant subspaces. It is not known if this result of Simonic [43] extends to the case of *complex* Hilbert spaces. Actually, Bishop-Phelps' theorem fails to extend to the complex case [33] and this may partly explain why the present section is so "real."

Our last corollary already has an hypercyclic flavor:

Corollary 2.8. *Let A be a unital subalgebra of $L(X)$, and let G be the connected component of I in A^{-1}. If $\overline{\text{span}}\{T(x);\ T \in G\} = X$ for all $x \in X\backslash\{0\}$, then in fact $\overline{\{T(x); T \in G\}} = X$ for all $x \in X\backslash\{0\}$.*

Proof. Our assumption implies that A is transitive. Connectedness shows that $\mathrm{Exp}(T) \in G$ for all $T \in A$, and thus Theorem 2.1 shows that X contains no non-trivial G-invariant closed subset. The conclusion follows. □

Remark. It should be stressed that Lomonosov's theorems, as well as the above Theorem 2.1, are mainly existence results, but they provide little information on the lattice of T-invariant subspaces. Hence, even if some smart topological argument happens to show that all operators in some given class have a non-trivial closed invariant subspace, this is by no means "the end of the story".

3. Hypercyclic operators

In this section, X denotes a separable Banach space, or more generally a separable Frechet space (that is, a locally convex metrizable complete space) over $\mathbb{R}$ or $\mathbb{C}$. An important example is the space $X = \mathcal{H}(\mathbb{C})$ of holomorphic functions from $\mathbb{C}$ to $\mathbb{C}$ equipped with the topology of compact convergence. If $M_z(f) = z \cdot f$ and $D(f) = f'$, one has $DM_z - M_z D = I$, and this is impossible for operators on normed spaces, hence $\mathcal{H}(\mathbb{C})$ cannot be made into a Banach space. If $x \in X$, we denote by

$$O_T(x) = \{T^n(x); n \geq 0\}$$

the orbit of x under T. We recall now a classical definition.

Definition 3.1. *An operator $T \in L(X)$ is hypercyclic if there exists $x \in X$ such that $O_T(x)$ is dense in X.*

Our first lemma asserts in particular that if the set

$$HC(T) = \{x \in X; \overline{O_T(x)} = X\}$$

is non-empty, then "almost every" $x \in X$ in the Baire category sense belongs to $HC(T)$.

Lemma 3.2. *$T \in L(X)$ is hypercyclic if and only if for all non-empty open sets U and V,*

$$U \cap \bigcap_{n \geq 1} T^{-n}(V) \neq \phi. \tag{3}$$

Proof. If T is hypercyclic and $x \in HC(T)$, then clearly $O_T(x) \subseteq HC(T)$ and thus $HC(T)$ is dense. Therefore (3) holds. If conversely (3) holds, we denote $\{V_k; k \geq 0\}$ a countable basis of the topology of X. By Baire's category lemma, (3) implies

$$\Omega = \bigcap_{k \geq 0} \left[\bigcup_{n \geq 1} T^{-n}(V_k) \right] \neq \phi$$

and it is clear that $\Omega = HC(T)$. □

We can now show an *hypercyclicity criterion* due to Carol Kitai [27]:

Lemma 3.3. *Let $T \in L(X)$ and assume that there is a sequence $(n_k) \subseteq \mathbb{N}$ increasing to ∞, dense subsets D and D' of X, and maps $S_{n_k} : D' \to X$ such that:*

(i) $T^{n_k}(x) \to 0$ for all $x \in D$
(ii) $S_{n_k}(y) \to 0$ for all $y \in D'$
(iii) $T^{n_k}S_{n_k}(y) \to y$ for all $y \in D'$.

Then T is hypercyclic.

Proof. Let U and V be non-empty open subsets. We will show that

$$T^{-n_k}(V) \cap U \neq \phi$$

for k large enough, and this suffices by Lemma 3.2. Pick $x \in U \cap D$ and $y \in V \cap D'$. We have

$$\lim_k T^{n_k}(x + S_{n_k}(y)) = \lim_k T^{n_k}(x) + T^{n_k}S_{n_k}(y) = y$$

therefore $T^{n_k}(x + S_{n_k}(y)) \in V$ for k large enough, while $(x + S_{n_k}(y)) \in U$ for k large enough. □

Remark. A simpler way to express the assumptions of Lemma 3.3. is: for any non-empty open sets U and V and any neighborhood W of 0, there exists $n \in \mathbb{N}$ such that $T^n(U) \cap W \neq \phi$ and $T^n(W) \cap V \neq \phi$. Other (non trivially) equivalent forms of this hypercyclicity criterion will be shown below.

The following special case (from [18]) of Lemma 3.3 is quite useful.

Lemma 3.4. *Let $T \in L(X)$ and denote*

$$X_T^+ = span\ \{\cup \ker(T - \lambda I);\ |\lambda| > 1\}$$

and

$$X_T^- = span\ \{\cup \ker(T - \lambda I);\ |\lambda| < 1\}.$$

If X_T^+ and X_T^- are dense subspaces of X, then T is hypercyclic.

Proof. Let $D = X_T^-$, $D' = X_T^+$ and $S = T^{-1} : D' \to D'$. If we let $S_n = S^n$, we can apply Lemma 3.3. □

Example. Let B be the backward shift on $\ell^2(\mathbb{N})$ such that $B((x_n)) = (x_{n+1})_{n\in\mathbb{N}}$, and let $|\lambda| > 1$. Then $T = \lambda B$ is hypercyclic [40] by Lemma 3.4: if $|\alpha| < 1$, $u_\alpha = (\alpha^n)_{n\geq 0}$ is an eigenvector of T and $T(u_\alpha) = \lambda\alpha u_\alpha$. It is easily shown that if Ω is any open set of $\{|z| < 1\}$, $\overline{\text{span}}\{u_\alpha; \alpha \in \Omega\} = \ell^2(\mathbb{N})$.

A similar method is used for showing a general result in [18]:

Theorem 3.5. *Let $T \in L(\mathcal{H}(\mathbb{C}^n))$ be an operator which commutes with the translation operators $\{\tau_z; z \in \mathbb{C}^n\}$ defined by $\tau_z(f)(u) = f(z + u)$, and $T \neq \lambda I$. Then T is hypercyclic.*

Such an operator T is commonly called a *pseudo-differential* operator, and we will see why in the proof below.

Proof. We work for convenience with $n = 1$. Let $\delta_0 \in \mathcal{H}(\mathbb{C})^*$ be such that $\delta_0(f) = f(0)$. We denote $\psi = T^*(\delta_0)$, and compute

$$\begin{aligned} Tf(z) &= \langle \delta_0, \tau_z T(f) \rangle = \langle \delta_0, T\tau_z(f) \rangle \\ &= \psi(\tau_z(f)). \end{aligned}$$

If we denote $P_k(u) = u^k$, Taylor's formula reads

$$\tau_z(f) = \sum_{k=0}^{\infty} \frac{f^{(k)}(z)}{k!} P_k$$

and thus

$$Tf(z) = \psi(\tau_z(f)) = \sum_{k=0}^{+\infty} \frac{f^{(k)}(z)}{k!} \psi(P_k).$$

If now $D(f) = f'$ is the differentiation operator, this formula reads

$$T(f) = \sum_{k=0}^{+\infty} \frac{\psi(P_k)}{k!} D^{(k)}(f)$$

and we have shown that there exists $F \in \mathcal{H}(\mathbb{C})$ such that $T = F(D)$.

Since $T \neq \lambda I$, F is not a constant function. For $\lambda \in \mathbb{C}$, let

$$e_\lambda(z) = e^{\lambda z} \in \mathcal{H}(\mathbb{C}).$$

We clearly have

$$T(e_\lambda) = F(\lambda)\, e_\lambda$$

and since $F \in \mathcal{H}(\mathbb{C})$ is not constant, the open sets

$$U^- = F^{-1}(\{|z| < 1\}); \; U^+ = F^{-1}(\{|z| > 1\})$$

are both non-empty. For applying Lemma 3.4, it suffices to show

Claim. If $U \subseteq \mathbb{C}$ is an open non-empty subset, then

$$E_U = \text{span}\,\{e_\lambda; \lambda \in U\}$$

is dense in $\mathcal{H}(\mathbb{C})$.

Indeed let $\phi \in \mathcal{H}(\mathbb{C})^*$ and define $G(\lambda) = \phi(e_\lambda)$. It is easily seen that $G \in \mathcal{H}(\mathbb{C})$ and that $G^{(k)}(0) = \phi(P_k)$ for all $k \geq 0$. If $\phi(e_\lambda) = 0$ for all $\lambda \in U$, we have $G(\lambda) = 0$ on U and thus $G = 0$ everywhere. It follows that $\phi(P_k) = 0$ for all $k \geq 0$ and thus $\phi = 0$. Hahn-Banach theorem then shows that claim. □

Theorem 3.5 extends early results ([11], [35]) for $T = \tau_1$ and $T = D$, which probably were the first examples given of hypercyclic operators.

Lemma 3.4 asserts that when T enjoys many eigenvectors corresponding to eigenvalues of modulus less than 1, and also of modulus greater than 1, then T is hypercyclic. It turns out that the conclusion still holds when T has many eigenvectors corresponding to eigenvalues of modulus 1, but this will require the use of quite different techniques.

Definition 3.6. *Let X be a Banach space over $\mathbb{C}$. We say that T has a perfectly spanning set of eigenvectors associated to eigenvalues of modulus 1 (in short, a p.s.s.) if there exists a continuous probability measure σ on $\mathbb{T}$ such that for all Borel subsets $A \subseteq \mathbb{T}(unit\ circle)$ with $\sigma(A) = 1$, one has*

$$X = \overline{span}\{\cup \ker(T - \lambda I);\ \lambda \in A\}.$$

In the statement above "σ continuous" means that $\sigma(\{\lambda\}) = 0$ for all $\lambda \in \mathbb{T}$.

Example. If $|\lambda| > 1,\ T = \lambda B$ has a p.s.s. with $d\sigma = d\theta/2\pi$ the Haar measure on $\mathbb{T}$.

We start with a qualitative theorem from [5].

Theorem 3.7. *If T has a p.s.s. then T is hypercyclic.*

Proof. We write down the proof in the case when σ is absolutely continuous, and with simple eigenvalues.

There exists a measurable map $E : \mathbb{T} \to X$ such that

$$\begin{cases} E(\lambda) & \neq 0 \Leftrightarrow \lambda \in \sigma_P(T) \cap \mathbb{T} \\ \|E(\lambda)\| & \leq 1 \text{ for all } \lambda \in \mathbb{T} \\ T(E(\lambda)) & = \lambda E(\lambda) \end{cases}$$

We define $K : L^2(\mathbb{T}, \sigma) \to X$ by

$$K(g) = \int_{\mathbb{T}} g(\lambda)\ E(\lambda)\ d\sigma(\lambda)$$

and a unitary operator $V : L^2(\mathbb{T}, \sigma) \to L^2(\mathbb{T}, \sigma)$ by

$$V(f)(z) = zf(z).$$

It is clear that

$$TK = KV. \tag{4}$$

We claim that $X_0 = K(L^2)$ is dense in X. Indeed if $x^* \in X_0^{\perp}$ we have

$$\langle x^*,\ E(\lambda)\rangle = 0$$

for σ-a.e. λ, and since T has a p.s.s. it follows that $x^* = 0$.

By (4), we have $T^n K = KV^n$ for all $n \geq 1$, hence if $x = K(g) \in X_0$ we can write

$$\begin{aligned} T^n(x) &= \int \lambda^n g(\lambda)\ E(\lambda)\ d\sigma(\lambda) \\ &= \hat{\mu}(-n) \end{aligned}$$

where $d\mu = gE\, d\sigma$ is an absolutely continuous X-valued measure. Now by absolute continuity

$$\lim_{|n|\to+\infty} \|\hat{\mu}(n)\| = 0 \tag{5}$$

and thus $\lim_{n\to\infty} T^n(x) = 0$ for all $x \in X_0$. We now define $S_n : X_0 \to X_0$ such that $T^n S_n = Id_{X_0}$ as follows. Given $x \in X_0$, and $g \in L^2(\mathbb{T}, \sigma)$ chosen such that $x = K(g)$, let

$$S_n(x) = \int_{\pi} \lambda^{-n}\, g(\lambda)\, E(\lambda)\, d\sigma(\lambda).$$

It follows again by (5) that

$$\lim_{n\to+\infty} \|S_n(x)\| = 0$$

and we apply Lemma 3.3 with $D = D' = X_0$. □

We now consider the case when $X = H$ is actually the complex separable Hilbert space. A *Gaussian measure m* on H is a centered probability measure such that if for all $x \in H$ we denote $f_x(y) = \langle y, x\rangle$ the corresponding linear form, then $f_x(m) = m \circ f_x^{-1}$ has a Gaussian distribution, or is δ_0 if m is supported by $\ker(f_x)$. We say that m is *non-degenerate* if it is supported by no closed hyperplane.

A Gaussian measure is determined by its *covariance operator S* defined by

$$\langle Sx, y\rangle = \int_H \langle x, z\rangle \overline{\langle y, z\rangle} dm(z)$$

S is a self-adjoint, positive, trace-class operator on H, and every such operator is a covariance operator. Moreover, m is non-degenerate if and only if S is one-to-one. We refer to [8] for more on this topic.

We now revisit the proof of Theorem 3.7, where the operator K is defined by

$$K : L^2(\mathbb{T}, \sigma) \to H$$

$$g \mapsto \int_{\pi} g(\lambda)\, E(\lambda)\, d\sigma(\lambda).$$

Since K has a square-summable kernel, it is a Hilbert-Schmidt operator and thus $S = KK^*$ is trace-class. Of course, $S = S^*$ and $S \geq 0$, thus S is the covariance operator of some non-degenerate Gaussian measure m. Indeed S is one-to-one, since $S(x) = KK^*(x) = 0$ implies

$$\langle x, KK^*(x)\rangle = \|K^*(x)\|^2 = 0$$

and K^* is one-to-one since K has dense range.

We now prove:

Lemma 3.8. *m is T-invariant.*

Proof. The covariance operator S' of $T(m) = m \circ T^{-1}$ is given by

$$\langle S'(x), y\rangle = \int_H \langle x, z\rangle \overline{\langle y, z\rangle}\, d(m \circ T^{-1}(z))$$
$$= \int \langle T^*(x), z\rangle \overline{\langle T^*(y), z\rangle}\, dm(z)$$

hence

$$\langle S'(x), y\rangle = \langle ST^*(x), y\rangle = \langle TST^*(x), y\rangle$$

and thus $S' = TST^*$. We now use (4) to conclude that

$$TST^* = TKK^*T^* = KVV^*K^* = KK^* = S$$

hence $S' = S$ and thus $T(m) = m$. □

We now recall a definition from ergodic theory: T is *strong-mixing* if for all A, B measurable

$$\lim_N |m(T^{-N}(A) \cap B) - m(A)\, m(B)| = 0$$

and it is *weak-mixing* if

$$\lim_N \frac{1}{N} \sum_{k=0}^{N-1} |m(T^{-k}(A) \cap B) - m(A)\, m(B)| = 0.$$

Such transformations are in particular ergodic. We can now state a quantitative improvement, shown in [6], of Theorem 3.7:

Theorem 3.9. *If $T \in L(H)$ has a p.s.s. relatively to some continuous measure σ, then there exists a non-degenerate T-invariant Gaussian measure m such that T is weak-mixing with respect to m. If σ is absolutely continuous, then T is actually strong-mixing.*

Proof. We first sketch the proof when σ is absolutely continuous. The reader should be warned that three "different" Hilbert spaces appear in this proof, namely H, $L^2(H, m)$ and $L^2(\mathbb{T}, \sigma)$. The corresponding scalar product will be denoted by $\langle \cdot, \cdot \rangle$ in every case but the reader is invited to check on which of these spaces it operates.

We denote

$$U_T(f) = f \circ T.$$

Since m is T-invariant, U_T is a unitary operator on $L^2(H, m)$. Showing that T is strong-mixing amounts to showing that for all $f, g \in L^2(H, m)$

$$\lim_{n\to\infty} \langle U_T^n(f), g\rangle = \langle f, 1\rangle \overline{\langle g, 1\rangle}.$$

If $f_x(z) = \langle z, x\rangle$, we have $f_x \in L^2(H, m)$ and $\langle f_x, 1\rangle = 0$.

Claim 3.10. For all x and y in H,

$$\lim_{k\to\infty} \langle U_T^k(f_x), f_y\rangle = 0.$$

Proof. We have

$$\langle U_T^k(f_x), f_y\rangle = \int_H \langle T^k(z), x\rangle\, \overline{\langle z, y\rangle}\, dm(z).$$

On the other hand,

$$\begin{aligned}\langle T S(y), x\rangle &= \langle S(y), T^{*k}(x)\rangle \\ &= \int_H \langle y, z\rangle\, \overline{\langle T^{*k}(x), z\rangle}\, dm(z) \\ &= \int_H \langle y, z\rangle\, \overline{\langle x, T^k(z)\rangle}\, dm(z)\end{aligned}$$

and therefore

$$\langle U_T^k(f_x), f_y\rangle = \langle T^k S(y), x\rangle. \tag{6}$$

Now, (4) implies that $T^k K = K V^k$ and since $S = K K^*$,

$$T^k S = K V^k K^*. \tag{7}$$

We conclude from (6) and (7) that

$$\langle U_T^k(f_x), f_y\rangle = \langle V^k K^*(y), K^*(x)\rangle.$$

Now, $K^*(x)$ and $K^*(y)$ belong to $L^2(\mathbb{T}, \sigma)$. We denote by

$$h = K^*(x) \cdot K^*(y)$$

their point-wise product. The above computation shows that

$$\langle U_T^k(f_x), f_y\rangle = \hat{h}(-k) \to 0.$$

□

To prove the theorem from the claim, one needs to go from the functions $\{f_x; x \in H\}$ to arbitrary functions $f \in L^2(H, m)$. This is done with a Stone-Weierstrauss type argument which uses the symmetric Fock spaces on H, and for which we refer to [6]. The case where σ is merely continuous is done along the same lines, but the convergence to 0 of the Fourier coefficients is replaced by Wiener's theorem stating that σ is continuous if and only if

$$\lim_{N\to\infty} \frac{1}{2N+1} \sum_{k=-N}^{N} |\hat{\sigma}(k)|^2 = 0$$

and this leads to the weak-mixing condition. □

We can now consider a quantitative form of hypercyclicity.

Definition 3.11. *An operator $T \in L(X)$ is called frequently hypercyclic if there is $x \in X$ such that for all $U \subseteq X$ open,*

$$\underline{\lim} \frac{1}{N} \#\{k \leq N;\, T^k(x) \in U\} > 0.$$

The set of such vectors x is denoted by $FHC(T)$.

With this terminology, we have [6]:

Theorem 3.12. *If $T \in L(H)$ has a p.s.s., then T is frequently hypercyclic.*

Proof. Since T is ergodic with respect to m, we have by Birkhoff's ergodic theorem that for m-almost every x,

$$\lim_{N \to +\infty} \frac{1}{N} \#\{k \leq N;\, T^k(x) \in U\} = m(U). \tag{8}$$

If $(U_\ell)_{\ell \geq 1}$ is a countable basis of the topology of H, it follows that for m-almost every x, (8) holds for all $\ell \geq 1$, and $m(U_\ell) > 0$ for all ℓ since m is non-degenerate. Therefore $m(FHC(T)) = 1$ and in particular T is frequently hypercyclic. □

Examples. Many natural hypercyclic operators are in fact frequently hypercyclic, and by using different techniques such examples are exhibited on Frechet spaces. For instance, every operator on $\mathcal{H}(\mathbb{C}^n)$ which satisfies the assumptions of Theorem 3.5 is in fact frequently hypercyclic. On the other hand, the backward shift on the Bergman space A_2 is hypercyclic but not frequently hypercyclic.

We refer to [6] for much more along these lines, and for a "frequent hypercyclicity criterion" in the spirit of Kitai's lemma 3.3.

4. Operators with prescribed behavior

We display in this Section some of Sophie Grivaux's works [21], where hypercyclicity will somehow meet invariant subspaces. We first recall a simple but very useful result ([12], [25], [9]):

Theorem 4.1. *Let X be a Banach space over $\mathbb{R}$ or $\mathbb{C}$, and $T \in L(X)$ an hypercyclic operator. If $x \in HC(T)$ and $P \in \mathbb{K}[\xi]\backslash\{0\}$, then $y = P(T)(x) \in HC(T)$. In particular, there exists a dense linear subspace M of X such that $M\backslash\{0\} \subseteq HC(T)$.*

Proof. We prove the result in the complex case.

Claim. For all $P \in \mathbb{K}[\xi]\backslash\{0\}$, $P(T)(X)$ is a dense subspace of X.

Proof of Claim. Indeed, if not there is $x_0^* \in X^*\backslash\{0\}$ such that $P(T^*)(x_0^*) = 0$, and thus there is a finite dimensional subspace V of X^* such that $T^*(V) \subset V$. But then $\sigma_p(T^*) \neq \phi$. Now if

$$T^*(x_1^*) = \lambda x_1^*$$

we have for all $x \in X$

$$\langle x_1^*, T^n(x)\rangle = \overline{\lambda}^n \langle x_1^*, x\rangle$$

and no geometric progression is dense in $\mathbb{C}$, hence $O_T(x)$ is not dense in X, contradicting T hypercyclic.

This proves the claim.

Continuing with the proof of the theorem:

Now if $y = P(T)(x)$ with $x \in HC(T)$ and $P \neq 0$, we have

$$O_T(y) = P(T)(O_T(x))$$

and thus $\overline{O_T(y)} = X$ since $\overline{O_T(x)} = X$ and $P(T)$ has dense range.

The space $M = \{P(T)(x);\ P \in \mathbb{K}[\xi]\}$ clearly works. □

Remark. The same proof works for frequently hypercyclic operators, replacing $HC(T)$ by $FHC(T)$.

The next theorem [21] shows that on *arbitrary* separable Banach spaces, an operator can be constructed which has a given orbit, up to the obvious necessary conditions.

We need the following lemma:

Lemma 4.2. *Let $\omega = \{u_n\}$ and $\omega' = \{w_n\}$ be two dense linearly independent sequences. For all $\epsilon > 0$, there exists $L \in L(X)^{-1}$ such that $\|1 - L\| < \epsilon$ and $L(\omega) = \omega'$.*

Proof. We sketch the proof and refer to [21] for the complete argument. Let (ϵ_n) be such that $\sum \epsilon_n < \epsilon < 1$. We construct by induction a sequence $(L_n) \subseteq L(X)^{-1}$ such that $L_0 = I$ and for all $n \geq 1$,

1. $\|L_n - K_{n-1}\| < \epsilon_n$.
2. $\exists\ p_n, q_n$ such that

$$L_n u_n = w_{p_n},\ L_n^{-1} w_n = u_{q_n}.$$

3. For all $n \geq 2$; $L_n = L_{n-1}$ on

$$\text{span}\,\{u_1, \dots, u_{n-1}, u_{q_1}, \dots u_{q_{n-1}}\}.$$

The operator $L = \lim_{n\to\infty} (L_n)$ works since

$$Lu_n = L_n u_n = w_{p_n}$$

and

$$Lu_{q_n} = L_n u_{q_n} = w_n.$$

□

Theorem 4.3. *Let X be a separable infinite dimensional Banach space. Let $\omega = \{u_n;\ n \geq 1\}$ be a dense linearly independent sequence in X. Then there exists $T \in L(X)$ such that $O_T(u_1) = \omega$.*

Proof. By a result of S. Ansari [1], given $\epsilon > 0$ there exists $T_0 \in L(X)$, an hypercyclic operator, such that $T_0 = I + N$ with $\|N\| < \epsilon$ and N a nuclear operator. To prove the theorem, we observe that $\overline{HC(T_0)} = X$, and thus by Lemma 4.2 there exists L with $\|1 - L\| < \epsilon$ and $x_0 \in HC(T_0)$ such that

$$L(u_1) = x_0$$

$$L(\omega) = O_{T_0}(x_0).$$

If we let now $T = L^{-1}T_0L$, we clearly have $O_T(u_1) = \omega$. □

Remark 1. Given $\epsilon > 0$, T can be chosen so that $\|T - I\| < 1 + \epsilon$. Observe moreover that $T = I + N$ with N a nuclear operator. It is not known whether there exists an infinte-dimensional Banach space X such that every operator on X is a nuclear perturbation of a scalar operator.

Remark 2. By a theorem of [13], an orbit which is somewhere dense (i.e. $\overline{O_T(x)}$ has a non-empty interior) is actually everywhere dense. Hence the denseness assumption from Theorem 4.2 is necessary.

The following Corollary answers a question of [24].

Corollary 4.4. *Let X be a separable infinite dimensional Banach space. Let $\omega \subseteq X$ be a linearly independent sequence, and pick $x \in \omega$. Then there exists $T \in L(X)$ such that $\omega \subseteq O_T(x)$.*

Proof. It is easy to show that ω is contained in a *dense* linearly independent sequence $\tilde{\omega}$, and Theorem 4.2 applies to $\tilde{\omega}$. □

We note in passing the following consequence of Lemma 4.3.

Corollary 4.5. *Let E and F be two normed spaces of countable algebraic dimension. If their completions are isomorphic, then E and F are isomorphic.*

We will now apply Theorem 4.1 and 4.2 together.

Corollary 4.6. *Let E be a normed space of countable algebraic dimension, and $x \in E\backslash\{0\}$. Then for any $\epsilon > 0$, there is $T \in L(E)$ such that $\|I - T\| < \epsilon$ and*

$$E = \{P(T)(x);\ P \in \mathbb{K}[\xi]\}.$$

Proof. There exists a dense basis ω of E containing x. If $T \in L(\overline{E})$ is such that $\|I - T\| < \epsilon$ and $O_T(x) = \omega$, we clearly have $T \in L(E)$ and $E = \mathbb{K}[T](x)$ as claimed. □

This will provide operators (on *non-complete* spaces!) with no proper closed invariant subset.

Corollary 4.7. *Let E be a normed space of countable algebraic dimension. Then there exists $T \in L(E)$ such that if F is a closed subset of E with $T(F) \subseteq F$, then $F = \{0\}$ or $F = E$.*

Proof. If T is given by Corollary 4.6, we have by Theorem 4.1 that for all $y = P(T)(x)$ with $P \neq 0$, $y \in HC(T)$, hence $\overline{O_T(y)} = E$. The conclusion follows since every $y \in E\backslash\{0\}$ can be written $y = P(T)(x)$. □

Note that C. Read [38] constructed an operator on the *complete* space $l_1(\mathbb{N})$ which satisfies the conclusion of Corollary 4.7.

Remark. Another way of stating Corollary 4.7 is to say that if $\|\cdot\|$ is an arbitrary norm on the space $\mathbb{K}[\xi]$ of polynomials, there exists a $\|\cdot\|$-continuous linear operator $T : \mathbb{K}[\xi] \to \mathbb{K}[\xi]$ such that for all $P \in \mathbb{K}[\xi]\backslash\{0\}$, the sequence $(T^n(P))_{n\geq 0}$ is $\|\cdot\|$-dense in $\mathbb{K}[\xi]$. It would be interesting to compute explicit examples of such operators T, for given norms $\|\cdot\|$, and to see whether some of them are reasonably simple.

5. About the hypercyclicity criterion

Until very recently, all known hypercyclic operators were satisfying the hypercyclicity criterion (Lemma 3.3). A combination of several works ([20], [10], [37]. [22]) shows that this condition is equivalent to several very natural ones. Let us say that an operator $T \in L(X)$ is hypercyclic with respect to a sequence $(n_k) \subseteq \mathbb{N}$ if there is $x \in X$ such that

$$\{T^{n_k}(x);\ k \in \mathbb{N}\}$$

is dense in X. When this happens, we also say that the sequence (T^{n_k}) is hypercyclic. With this notation, the following holds.

Theorem 5.1. *Let X be a separable Banach space, and $T \in L(X)$ be an hypercyclic operator. Then the following assertions are equivalent:*

(i) T satisfies the hypercyclicity criterion.
(ii) $T \oplus T \in L(X \oplus X)$ is hypercyclic.
(iii) There exists a strictly increasing sequence (n_k) such that T is hypercyclic with respect to any subsequence of (n_k).
(iv) For every increasing sequence (n_k) such that $\sup\{n_{k+1} - n_k;\ k \geq 0\} = M < \infty$, *the sequence (T^{n_k}) is hpercyclic.*
(v) Same as (iv), but with $M = 2$.

It is known since the Fall of 2006 that there are hypercyclic operators which fail to satisfy the above equivalent conditions [39]. This construction was followed by [7] where the complexity of the argument is somehow transferred from the norm to the operator, and where it is shown that certain hypercyclic operators on the Hilbert space fail the hypercyclicity criterion. Note however that every *frequently* hypercyclic operator satisfies it [23]. This direction of research is far from exhausted, and in particular it would be interesting to find natural hypercyclic operators on the Hilbert space, with a simple analytic definition, which would fail the hypercyclicity criterion.

6. A tribute to Gustave Choquet

Professor Gustave Choquet died on November 14, 2006. He was born in 1915 in Solesmes, France. A student of Ecole Normale Supérieure, G. Choquet ranked first in Agrégation in 1937 (his competitor and friend Laurent Schwartz was only second best in this same contest) and after studying under the direction of Arnaud Denjoy, he defended his Thèse (Ph.D) in 1946. Among the main results of his memoir stands the characterization, up to an homeomorphism on the domain, of derivatives of real-valued functions of a real variable as those functions which satisfy the intermediate value property and are pointwise limits of a sequence of continuous functions. He held positions in the French Institute in Warsaw, and later Grenoble, before obtaining a position in Paris where he taught until his retirement in 1984. He was a member of the French Academy of Sciences since 1976.

Among the many important contributions of Gustave Choquet, let us mention two major items. In his theory of capacities, he completely renews potential theory through an abstract but very powerful notion of "capacity", a set functional which formalizes the corresponding physical notion. On the other hand, integral representation theory (which is sometimes simply called Choquet theory) uses very simple geometric ideas for proving representation theorems which refine Krein-Milman theorem and show that any point of a compact convex set is a mixture of extreme points. A simple proof of one of these theorems is displayed below.

Gustave Choquet devoted a fair amount of his time to teaching, and to a personnal analysis of the best way to teach mathematics. He certainly is a member of the Bourbaki generation, but he never was a member of the Bourbaki group. Let me recall for the younger readers that Bourbaki denotes a "secret society" of mathematicians (most of them French) who originally intended to reorganize mathematics as they were in the twentieth century in a logical order, and (so to speak) to promote the idea that teaching should be delivered in this same logical order. All members of Bourbaki are mathematicians of first magnitude, but the tradition wants that no member ever claims to be part of this group nor mentions it on its Vita, although contributing to Bourbaki means a fair amount of work. Bourbaki produced its "Eléments de Mathématique" and it is not by chance that this title is reminiscent of Euclid's famous work.

A high school student in Paris in the sixties, when Bourbaki was at the peak of its power, the author was for instance taught notions such as commutativity and associativity when he was 10. He can therefore testify how difficult it is to grasp at this stage the relevance of such concepts, which look like empty talking when only trivial examples are available. Gustave Choquet vigorously opposed such an abstract way of teaching mathematics, and stressed instead the importance of examples, observations and links with physics and other sciences. With all due respect to Bourbaki, I dare to say he was right.

I was privileged to be a student of Gustave Choquet. His geometric intuition and his illuminating ability to simplify problems brought with them the comforting feeling that mathematics are simpler than what they look at first sight. I remember his comment after attending a very technical lecture: "it's complicated, therefore it is not mathematics". He was joking of course, but such jokes are significant. And reassuring for beginners who are afraid to get lost in the wilderness of mathematical knowledge, or in the boundless complexity of technical details.

What kind of advisor was he? A director who did not want to lead his students on a narrow road, but instead had full respect of their freedom of choice. Mr Choquet was not assigning specific problems or tasks, and what mattered for him was the weekly seminar and the questions which might, or not, arise from the lecture or the conversations between the little crowd of regular participants. In this simple way, Mr Choquet opened for us the way to a pleasant accomplishment of our own mathematical inclinations. And one more thing: he never played the mischievous game of publicly comparing his students. My immediate surrounding was twinkling with young stars. What a relief it was to hear from him this simple piece of advice: "don't try to become the conformal copy of somebody else".

Thanks to Professor Choquet, I learned that truth, simplicity and beauty often coincide. I learned to taste and like, to be neither master nor slave to mathematics. Thanks to him, I have been able to sing my little tune in the great concert of mathematics, to ask many questions, to solve a few of them. What else should I say? Merci, monsieur Choquet.

Let us now outline some results from integral representation theory.

The frame of this theory is a subset (call it K) of a locally convex Hausdorff topological vector space which is convex and compact for the induced topology. A point $e \in K$ is called an extreme point if the equation

$$e = \frac{x+y}{2}$$

with x and y in K implies that $x = y = e$. The set of extreme points is denoted by $\mathrm{Ext}(K)$.

If K is a compact convex set, the set $\mathrm{Ext}(K)$ is not necessarily closed. In fact, there is $K \subseteq \mathbb{R}^3$ such that $\mathrm{Ext}(K)$ is not closed. Another example is provided by the unit ball of the Hilbert space equipped with its weak topology, for which the set $\mathrm{Ext}(K)$ is clearly dense in K.

However, when K is metrizable, $\mathrm{Ext}(K)$ is a G_δ-set (that is, a countable intersection of open sets). Since if d is a metric which defines the topology of K, then

$$\mathrm{Ext}\,(K) = K \Big\backslash \left(\bigcup_{n \geq 1} K_n \right)$$

where

$$K_n = \left\{ \frac{x+y}{2};\ (x, y) \in K^2,\ d(x, y) \geq \frac{1}{n} \right\}$$

is a compact subset of K. In general, $\mathrm{Ext}(K)$ can be quite complicated but it is always an "α-favorable" topological space, and thus in particular a Baire space. This simple result of G. Choquet ranks him among the founding fathers of the modern theory of topological games.

Let us now state and prove the Bishop-Choquet-de Leeuw integral representation theorem. This result leads to a large number of representation theorems for elements of compact convex sets, which are often defined by inequalities, through the extreme points of these convex sets, which are often defined by the corresponding equations and therefore enjoy algebraic properties. One example among many: the extreme points of the compact convex set of normalized positive definite functions on a compact Abelian group are the characters of the group, and the integral representation theorem is in this case Bochner's theorem. We refer to [14] for more examples, and an exposition of the whole theory.

Theorem 6.1. *Let K be a convex compact metrizable set. For any $x \in K$, there is a probability measure μ on K such that $\mu(Ext\,(K)) = 1$, and $x = b(\mu)$ is the barycenter of μ; that is*

$$f(x) = \int_{Ext\,(K)} f(e)\, d\mu(e)$$

for every affine continuous function f on K.

Proof. Since K is metrizable, there is a sequence $\{f_n;\, n \geq 1\}$ of affine continuous functions on K which separate the points of K, and such that

$$\sup_n \|f_n\|_\infty < \infty.$$

If we let

$$\varphi(x) = \sum_{n=1}^{\infty} 2^{-n} f_n(x)^2$$

then φ is a *strictly* convex continuous function on K.

We denote

$$\hat{\varphi} = \inf\{h;\, h \geq \varphi,\, h \text{ affine continuous}\}$$

the function $\hat{\varphi}$ is the concave upper semi-continuous envelope of φ, and its subgraph

$$g^-(\hat{\varphi}) = \{(x,t) \in K \times \mathbb{R};\ t \leq \hat{\varphi}(x)\}$$

is easily seen to satisfy

$$g^-(\hat{\varphi}) = \overline{\text{conv}}[g^-(\varphi)].$$

Therefore, given $x \in K$, there exists $\mu \in \mathcal{P}(K)$ which satisfy the following two conditions:

(i) $x = b(\mu)$.
(ii) $\hat{\varphi}(x) = \int \varphi\, d\mu$.

We claim that $\mu(\text{Ext}(K)) = 1$ and thus that μ satisfies the conclusion of the theorem. Indeed, since $\hat{\varphi}$ is concave u.s.c. we have

$$\hat{\varphi}(x) \leq \int \varphi\, d\mu \leq \int \hat{\varphi}\, d\mu \leq \hat{\varphi}(x).$$

Therefore

$$\int \varphi \, d\mu = \int \hat{\varphi} \, d\mu$$

and since $\varphi \leq \hat{\varphi}$, it follows that $\varphi = \hat{\varphi}$, μ-almost everywhere. But if $x \notin \mathrm{Ext}\,(K)$, we write

$$x = \frac{y+z}{2}$$

with $y \neq z$ in K, and thus

$$\begin{aligned} \varphi(x) &< \frac{1}{2}\,[\varphi(y) + \varphi(z)] \\ &\leq \frac{1}{2}\,[\hat{\varphi}(y) + \hat{\varphi}(z)] \leq \hat{\varphi}(x) \end{aligned}$$

and $\varphi(x) < \hat{\varphi}(x)$ if $x \notin \mathrm{Ext}\,(K)$.

We have shown that μ-almost every $x \in K$ is an extreme point, or in other words that $\mu(\mathrm{Ext}\,(K)) = 1$. □

Gustave Choquet elucidated as well the important problem of the uniqueness of the integral representation. A convex set which satisfy uniqueness is called a Choquet simplex. And this is a nice case where geometric intuition helps: a triangle is the simplest case of a simplex, and the intersection of two similar triangles is again a similar triangle. This leads to the idea that a simplex is a compact convex set such that the generated cone is a lattice cone: this is Choquet's definition, and it works.

Let us observe that in the spirit of the above proof, it is possible to give a simple proof [19] of the existence of the Haar measure for metrizable compact groups. This is not more than an exercise, but let me respectfully call it a student's tribute to Gustave Choquet.

Let G be such a group, and denote

$$\tau_g(h) = gh$$

for all $g \in G$. We wish to show the existence of a probability measure $\mu \in P(G)$ such that

$$\tau_g(\mu) = \mu \circ \tau_g^{-1} = \mu$$

for all $g \in G$. Consider as above a strictly convex continuous function

$$\varphi : P(G) \to \mathbb{R}$$

and define for all $\mu \in P(G)$

$$\tilde{\varphi}(\mu) = \sup\{\varphi(\tau_g(\mu));\ g \in G\}.$$

It is clear that $\tilde{\varphi}$ is lower semi-continuous and convex. We claim that is even *strictly* convex. Indeed, let

$$\lambda = \frac{\mu + \nu}{2}$$

with $\mu \neq \nu$ in $P(G)$.

By compactness, there is $g_0 \in G$ such that

$$\tilde{\varphi}(\lambda) = \varphi(\tau_{g_0}(\lambda))$$

and then we have

$$\begin{aligned}\tilde{\varphi}(\lambda) &= \varphi\left(\frac{1}{2}\left(\tau_{g_0}(\nu) + \tau_{g_0}(\mu)\right)\right)\\ &< \frac{1}{2}\left(\varphi(\tau_{g_0}(\nu)) + \varphi(\tau_{g_0}(\mu))\right)\\ &\leq \frac{1}{2}\left[\tilde{\varphi}(\nu) + \tilde{\varphi}(\mu)\right]\end{aligned}$$

and thus $\tilde{\varphi}$ is strictly convex. Since it is also l.s.c., it attains its minimum at a *unique* $\mu \in P(K)$.

But since one clearly has

$$\tilde{\varphi} \circ \tau_g = \tilde{\varphi}$$

for all $g \in G$, it follows that

$$\tau_g(\mu) = \mu$$

for all $g \in G$. □

I am very grateful to Professor T. S. S. R. K. Rao for the perfect organization of this gathering, and for the very open and exciting atmosphere of the conference. Thanks are also due to all the Indian colleagues and friends who kindly took care of their visitors, and successfully made sure that their stay was both a very nice mathematical and a very nice personal experience.

References

[1] S. I. Ansari, Existence of hypercyclic operators on topological vector spaces, *J. Funct. Anal.* **148** (1997) 384–390.

[2] A. Atzmon, The existence of translation invariant subspaces of symmetric self-adjoint sequence spaces on Z, *J. Funct. Anal.* **178** (2000) 372–380.

[3] A. Atzmon and G. Godefroy, An application of the smooth variational principle to the existence of nontrivial invariant subspaces, *C. R. Acad. Sci. Paris Sér. I Math.* **332** (2001) 151–156.

[4] A. Atzmon, G. Godefroy and N. J. Kalton, Invariant subspaces and the exponential map, *Positivity* **8** (2004) 101–107.

[5] F. Bayart and S. Grivaux, Hypercyclicity and unimodular point spectrum, *J. Funct. Anal.* **226** (2005) 281–300.

[6] F. Bayart and S. Grivaux, Frequently hypercyclic operators, *Trans. Amer. Math. Soc.* **358** (2006) 5083–5117.

[7] F. Bayart and E. Matheron, Construction d'operateurs hypercycliques ne verifiant pas le critere d'hypercyclicite, (French) *Construction of hypercyclic operators failing the hypercyclicity criterion, C. R. Math. Acad. Sci. Paris* **344** (2007) 231–234.

[8] Y. Benyamini and J. Lindenstrauss, Geometric nonlinear functional analysis, Vol. 1. American Mathematical Society Colloquium Publications, 48. American Mathematical Society, Providence, RI (2000) xii+488 pp.

[9] J. P. Bès, Invariant manifolds of hypercyclic vectors for the real scalar case, *Proc. Amer. Math. Soc.* **127** (1999) 1801–1804.

[10] J. P. Bès and A. Peris, Hereditarily hypercyclic operators, *J. Funct. Anal.* **167** (1999) 94–112.

[11] G. D. Birkhoff, Demonstration d'un theoreme elementaire sur les fonctions entieres, *Note aux CRAS Paris* **189** (1929) 473–475.

[12] P. S. Bourdon, Invariant manifolds of hypercyclic vectors, *Proc. Amer. Math. Soc.* **118** (1993) 845–847.

[13] P. S. Bourdon and N. S. Feldman, Somewhere dense orbits are everywhere dense, *Indiana Univ. Math. J.* **52** (2003) 811–819.

[14] G. Choquet, Lectures on analysis. Vol. I, II and III, Edited by J. Marsden, T. Lance and S. Gelbart W. A. Benjamin, Inc., New York-Amsterdam (1969).

[15] R. Deville, G. Godefroy and V. Zizler, A smooth variational principle with applications to Hamilton-Jacobi equations in infinite dimensions, *J. Funct. Anal.* **111** (1993) 197–212.

[16] P. Enflo, On the invariant subspace problem for Banach spaces, *Acta Math.* **158** (1987) 213–313.

[17] P. Enflo and V. Lomonosov, Some aspects of the invariant subspace problem, Handbook of the geometry of Banach spaces, Vol. I, 533–559, North-Holland, Amsterdam (2001).

[18] G. Godefroy and J. H. Shapiro, Operators with dense, invariant, cyclic vector manifolds, *J. Funct. Anal.* **98** (1991) 229–269.

[19] G. Godefroy and D. Li, Strictly convex functions on compact convex sets and applications, Functional analysis, 182–192, Narosa, New Delhi (1998).

[20] L. Bernal-Gonzalez and K-G. Grosse-Erdmann, The hypercyclicity criterion for sequences of operators, *Studia Math.* **157** (2003), 17–32.

[21] S. Grivaux, Construction of operators with prescribed behaviour, *Arch. Math. (Basel)* **81** (2003) 291–299.

[22] S. Grivaux, Hypercyclic operators, mixing operators, and the bounded steps problem, *J. Operator Theory* **54** (2005) 147–168.

[23] K-G. Grosse-Erdmann and A. Peris, Frequently dense orbits, *C. R. Math. Acad. Sci. Paris* **341** (2005) 123–128.

[24] I. Halperin, C. Kitai and P. Rosenthal, On orbits of linear operators, *J. London Math. Soc.* **31** (1985) 561–565.

[25] D. A. Herrero, Limits of hypercyclic and supercyclic operators, *J. Funct. Anal.* **99** (1991) 179–190.

[26] N. D. Hooker, Lomonosov's hyperinvariant subspace theorem for real spaces, *Math. Proc. Cambridge Philos. Soc.* **89** (1981) 129–133.

[27] C. Kitai, Invariant closed sets for linear operators, PhD Thesis, University of Toronto (1982).

[28] M. G. Krein and M. A. Rutman, Linear operators leaving invariant a cone in a Banach space, (Russian) *Uspehi Matem. Nauk (N. S.)* **3** no. 1(23) (1948) 3–95.

[29] M. Lindstrom and G. Schluchtermann, Lomonosov's techniques and Burnside's theorem, *Canad. Math. Bull.* **43** (2000) 87–89.

[30] V. Lomonosov, Invariant subspaces of the family of operators that commute with a completely continuous operator, (Russian) *Funkcional. Anal. i Prilozen.* **7** (1973) 55–56.
[31] V. Lomonosov, An extension of Burnside's theorem to infinite-dimensional spaces, *Israel J. Math.* **75** (1991) 329–339.
[32] V. Lomonosov, Positive functionals on general operator algebras, *J. Math. Anal. Appl.* **245** (2000) 221–224.
[33] V. Lomonosov, A counterexample to the Bishop-Phelps theorem in complex spaces, *Israel J. Math.* **115** (2000) 25–28.
[34] V. Lomonosov and P. Rosenthal, The simplest proof of Burnside's theorem on matrix algebras, *Linear Algebra Appl.* **383** (2004) 45–47.
[35] G. R. MacLane, Sequences of derivatives and normal families, *J. Analyse Math.* **2** (1952) 72–87.
[36] C. Pearcy and A. L. Shields, A survey of the Lomonosov technique in the theory of invariant subspaces, *Topics in operator theory*, pp. 219–229. Math. Surveys, No. 13, Amer. Math. Soc., Providence, R.I. (1974).
[37] A. Peris and L. Saldivia, Syndetically hypercyclic operators, *Integral Equations Operator Theory* **51** (2005) 275–281.
[38] C. J. Read, A solution to the invariant subspace problem, *Bull. London Math. Soc.* **16** (1984) 337–401.
[39] M. De La Rosa and C. J. Read, *An hypercyclic operator whose direct sum is not hypercyclic*, Preprint (2006).
[40] S. Rolewicz, On orbits of elements, *Studia Math.* **32** (1969) 17–22.
[41] W. Rudin, *Functional analysis. Second edition. International Series in Pure and Applied Mathematics.* McGraw-Hill, Inc., New York (1991).
[42] T. Schlumprecht and V. G. Troitsky, On quasi-affine transforms of Read's operator, *Proc. Amer. Math. Soc.* **131** (2003) 1405–1413 (electronic).
[43] A. Simonic, An extension of Lomonosov's techniques to non-compact operators, *Trans. Amer. Math. Soc.* **348** (1996) 975–995.
[44] G. Sirotkin, A version of the Lomonosov invariant subspace theorem for real Banach spaces, *Indiana Univ. Math. J.* **54** (2005) 257–262.
[45] V. G. Troitsky, Lomonosov's theorem cannot be extended to chains of four operators, *Proc. Amer. Math. Soc.* **128** (2000) 521–525.

Proc. Int. Conf. – Lectures on Operator Theory No. 8, 2009, pp. 51–85.

An Introduction to E_0-semigroups

Geoffrey Price

United States Naval Academy, Annapolis, Maryland U.S.A.
e-mail: glp@usna.edu

1. C^*-algebras and von Neumann algebras

Preliminaries

The study of operator algebras may be said to have begun in the 1930s with the papers of F. J. Murray and J. von Neumann on Rings of Operators. Another early and major contributor is M. H. Stone, whose book on Linear Transformations in Hilbert Space is of fundamental importance to the subject, [Sto]. Rings of operators are now known as von Neumann algebras. A little later came the theory of C^*-algebras which we discuss briefly below. They may be thought of as Banach algebras over the complex numbers $\mathbb{C}$ which satisfy a special norm condition. C^*-algebras have arisen as special types of limits of algebras of local observables of a quantum mechanical system. Although C^*-algebras are still applied in quantum mechanics and quantum statistics they have proved useful in other areas of mathematics, including knot theory, number theory, and additional areas of physics.

The purpose of these lectures is to give an introduction to some of the mathematics which comes up in the study of operator algebras and mathematical physics. The goal is to reach a point where we can discuss some of the current work going on in C^*-dynamical systems. In particular we will introduce the notion of E_0-semigroups. This is a subject which began with the work of R. T. Powers in the 1980s. The origin of this study may be thought of as arising from a question raised by Powers at that time: It is known, from a famous result of Wigner in the 1930s, how to describe all continuous one-parameter groups of $*$-automorphisms on $\mathcal{B}(\mathcal{H})$. These may be thought of, incidentally, as the (reversible) time evolution of the observables of a physical system. How then, can one describe the non-reversible time evolutions of a physical system?

In these notes I hope to outline some of the important ideas that one encounters in the process of addressing this question. These ideas will take us through some of the results in the theory of unbounded symmetric operators on a Hilbert space and the

#Supported in part by U.S. National Science Foundation Grants DMS-0400841 and DMS-0700469 and by a travel grant from the National Board for Higher Mathematics of India.

related theory of unbounded derivations on C^*-algebras and von Neumann algebras. At the end we will be able to see how Powers looked at this problem originally and we will discuss some of his early results. We can then hint at some of the beautiful work that has emerged since, including the important contributions of Powers, William Arveson, and of B. V. R. Bhat, and others.

The author gratefully acknowledges the support of the National Board for Higher Mathematics and the kind hospitality of Dr. B. V. Rajarama Bhat and Dr. T. S. S. R. K. Rao, the convener of the workshop, held at the Indian Statistical Institute, Bangalore.

Introduction to C^*-algebras

A C^*-algebra $\mathfrak{A}$ is a Banach $*$-algebra which has the norm condition $\|A^*A\| = \|A\|^2$ for all elements A in the algebra. As we will see the algebra $\mathcal{B}(\mathcal{H})$ of all bounded linear operators on a Hilbert space $\mathcal{H}$ is an example. To introduce some of the ideas that we will encounter in these lectures we present a small portfolio of examples below.

Example 1.1. Consider the commutative algebra $C(X)$ consisting of the continuous complex-valued functions on a compact Hausdorff space X, equipped with the supremum norm

$$\|f\| =\underset{x\in X}{\rightarrow} \sup|f(x)|.$$

The $*$-operation is $f^*(x) = \bar{f}(x)$. Observe that $\|f\|^2 = \|f\bar{f}\| =\underset{x\in X}{\rightarrow} \sup|f(x)|^2$.

Example 1.2. For any integer $n > 1$ let $\mathfrak{A}$ be the algebra $M_n(\mathbb{C})$ of $n \times n$ matrices with entries in the complex numbers $\mathbb{C}$. Set $A^* = \bar{A}^t$, the complex conjugate of the transpose of A.

Let $\mathbb{C}^n$ be the vector space over $\mathbb{C}$ of column vectors of length n. We endow $\mathbb{C}^n$ with the usual inner product $(f, g) = \sum_{j=1}^n f_j\bar{g}_j$, (where $f = (f_1, f_2, \dots, f_n)^t$, $g = (g_1, g_2, \dots, g_n)^t$ and t denotes transpose), and norm $\|f\| = (f, f)^{\frac{1}{2}}$. Then a straightforward calculation shows that

$$(Af, g) = (f, A^*g)$$

and therefore A^*A is a *positive* operator, or matrix, in the sense that for any $f \in \mathbb{C}^n$, $(A^*Af, f) \geq 0$. We can and often do view $\mathfrak{A}$ as the algebra of linear transformations on $\mathbb{C}^n$, acting on the column vectors $f \in \mathbb{C}$ by left multiplication, i.e. if $A \in \mathfrak{A}$ and f is in $\mathbb{C}^n$ then $Af \in \mathbb{C}^n$. Let's verify that the norm

$$\|A\| = \max_{\|f\|=1} \|Af\|$$

is a C^*-algebra norm, i.e., that $\|A\|^2 = \|A^*A\|$. To see this we will use the so-called polar decomposition of a matrix A, [A4]. Since the matrix A^*A is diagonalizable we know that it has a unique positive square root which we will denote Δ. We have

$$\|Af\|^2 = (Af, Af) = (A^*Af, f) = (\Delta^2 f, f) = (\Delta f, \Delta f).$$

This suggests that we consider the mapping W on $\mathbb{C}^n$ defined as follows: for $g = \Delta f$ in $Ran(\Delta)$, the range of Δ, set $Wg = Af$ and for $g \in Ran(\Delta)^\perp = \ker(\Delta) = \ker(A)$, set $Wg = 0$. Then $W\Delta = A$ and W is an isometry on $Ran(\Delta)$, so

$$\|A\| = \sup_{\|f\|=1} \|W\Delta f\| = \sup_{\|f\|=1} \|\Delta f\| = \|\Delta\|.$$

Since Δ is a positive (and therefore diagonalizable) matrix it is clear that $\|\Delta\|^2 = \|\Delta^2\| = \|A^*A\|$ so $\|A\|^2 = \|A^*A\|$ and we are done.

Example 1.3. Let $\mathcal{H}$ be a separable Hilbert space and let $\mathcal{B}(\mathcal{H})$ be the algebra of all bounded linear transformations on $\mathcal{H}$. Then for each $A \in \mathcal{B}(\mathcal{H})$ there is a unique operator A^* which satisfies

$$(Af, g) = (f, A^*g)$$

for all $f, g \in \mathcal{H}$. If $\dim(\mathcal{H}) < \infty$ then $\mathcal{B}(\mathcal{H})$ is isomorphic to the algebra of $n \times n$ matrices and we have shown above that this is a C^*-algebra. If $\dim(\mathcal{H}) = \infty$ choose an orthonormal basis $\{f_i : i \in \mathbb{N}\}$ for $\mathcal{H}$. Then for each $i \in \mathbb{N}$ there are coefficients a_{ji}, $j \in \mathbb{N}$ such that $Af_i = \sum_{j=1}^{\infty} a_{ji} f_j$, suggesting that we may view the linear operator A as an infinite matrix

$$\begin{pmatrix} a_{11} & a_{12} & \cdots \\ a_{21} & a_{22} & \cdots \\ \vdots & \vdots & \end{pmatrix}.$$

Of course not every infinite matrix corresponds to a bounded linear operator on $\mathcal{H}$. Note that each column of the matrix, for example, must have coefficients in $\ell^2(\mathbb{N})$.

A finite rank operator is one that with respect to some orthonormal basis has the form

$$\begin{pmatrix} a_{11} & a_{12} & \ldots & a_{1n} & 0 & \ldots \\ a_{21} & a_{22} & \ldots & a_{2n} & 0 & \ldots \\ \vdots & \ldots & \ldots & \ldots & \ldots & \\ a_{n1} & a_{n2} & \ldots & a_{nn} & 0 & \ldots \\ 0 & 0 & 0 & 0 & 0 & \ldots \\ 0 & 0 & 0 & 0 & 0 & \ldots \\ \vdots & \ldots & & & & \end{pmatrix}$$

for some n. If one takes the norm closure of the finite rank operators one obtains a proper C^*-algebra of $\mathcal{B}(\mathcal{H})$ which is called the algebra of compact operators, $\mathcal{C}(\mathcal{H})$, which is actually a two-sided ideal of $\mathcal{B}(\mathcal{H})$.

Example 1.4. The CAR-algebra. *CAR* stands for Canonical Anticommutation Relations. The CAR-algebra is an important algebra in numerous applications of mathematical physics. It is an algebra which is isomorphic to the so-called Clifford algebra. It is the underlying algebra in the Ising model and other important physical models, [BR].

We can introduce ourselves to this algebra by thinking about the problem of how to embed the algebra $M_2(\mathbb{C})$ of 2×2 matrices into $M_4(\mathbb{C})$. One way is to embed a 2×2 matrix $A = \left(\begin{smallmatrix} a & b \\ c & d \end{smallmatrix}\right)$ into a 4×4 matrix by inserting it into the top left corner, i.e.,

$$A \to \begin{pmatrix} a & b & 0 & 0 \\ c & d & 0 & 0 \\ 0 & 0 & 0 & 0 \\ 0 & 0 & 0 & 0 \end{pmatrix}.$$

Note that with this embedding the 2×2 identity matrix I_2 does not embed into the 4×4 identity matrix. Suppose that is what we want, though. Then consider the embedding which takes the 2×2 matrix A and embeds it as

$$\begin{pmatrix} A & \\ & A \end{pmatrix}.$$

Note that with this embedding the identity matrix I_2 embeds into the identity I_4. This is what we want.

The CAR algebra is sometimes written symbolically as a sequence of embeddings

$$I \to M_2(\mathbb{C}) \to M_4(\mathbb{C}) \to M_8(\mathbb{C}) \to \cdots$$

or even more cryptically as

$$1 \to 2 \to 4 \to 8 \to \cdots$$

where the numbers $1, 2, 4, 8, \ldots$ represent matrices of the given size. Here is how we might think of this. Recall that for a separable Hilbert space $\mathcal{H}$ the algebra $\mathcal{B}(\mathcal{H})$ consists of certain matrices of the form

$$\begin{pmatrix} a_{11} & a_{12} & \cdots \\ a_{21} & a_{22} & \cdots \\ \vdots & \vdots & \end{pmatrix}.$$

The algebra of 2×2 matrices A may be represented as one of these, where we just repeat the matrix A infinitely often along the main diagonal (and with 0s everywhere else), i.e.,

$$A \hookrightarrow \begin{pmatrix} A & & & \\ & A & & \\ & & A & \\ & & & \ddots \end{pmatrix}.$$

The 4×4 matrices embed in the same way, i.e., if $B = \left(\begin{smallmatrix} B_{11} & B_{12} \\ B_{21} & B_{22} \end{smallmatrix}\right)$ is a 4×4 matrix consisting of the 2×2 B_{ij}s then

$$B \hookrightarrow \begin{pmatrix} B_{11} & B_{12} & & & \\ B_{21} & B_{22} & & & \\ & & B_{11} & B_{12} & \\ & & B_{21} & B_{22} & \\ & & & & \ddots \end{pmatrix}.$$

We continue this procedure forever (embedding the 8×8 matrices, then the 16×16 matrices, etc.), obtaining an algebra which is contained in $\mathcal{B}(\mathcal{H})$. Then we take the norm closure of this $*$-algebra, which is the CAR algebra.

Intuitively one can see that the CAR algebra is not all of $\mathcal{B}(\mathcal{H})$ by observing that the finite rank matrices in $\mathcal{B}(\mathcal{H})$ (see the previous example) can not be approximated by the sort of embedding we are making of the $2^n \times 2^n$ matrices into $\mathcal{B}(\mathcal{H})$. In fact it is true that the only compact operator in the CAR algebra is 0.

Here is another way to view the sequence of embeddings we are making to form the CAR algebra. We know that the matrix algebra $M_4(\mathbb{C})$ may be identified with the tensor product $M_2(\mathbb{C}) \otimes M_2(\mathbb{C})$ and so we embed $A \in M_2(\mathbb{C})$ into $M_4(\mathbb{C})$ by sending A to $A \otimes I_2$. More generally, for any integer $n > 2$ the algebra of $2^n \times 2^n$ matrices is isomorphic to the tensor product

$$\underbrace{M_2(\mathbb{C}) \otimes M_2(\mathbb{C}) \otimes \cdots \otimes M_2(\mathbb{C})}_{n \text{ times}}$$

and the algebra of 2×2 matrices may be embedded into this tensor product in any one of the n positions by sending $A \in M_2(\mathbb{C})$ to $I \otimes I \otimes \cdots I \otimes A \otimes I \otimes \cdots \otimes I$, where I represents the 2×2 identity matrix.

In fact we can get a little looser with the notation and think of the $2^n \times 2^n$ matrices as the tensor product

$$\underbrace{M_2(\mathbb{C}) \otimes M_2(\mathbb{C}) \otimes \cdots \otimes M_2(\mathbb{C})}_{n \text{ times}} \otimes \underbrace{I \otimes I \otimes I \cdots}_{\text{infinitely many}}$$

where we tack on the infinite tensor product of 2×2 identity matrices just for good measure.

Finally let's give the algebra $M_2(\mathbb{C})$ infinitely different names, i.e., let $\mathcal{A}_1, \mathcal{A}_2, \ldots$ all be copies of the algebra of 2×2 matrices. Then we can imagine positioning the jth copy $\mathcal{A}_j$ into the jth spot of a tensor product otherwise consisting of 2×2 identity matrices, i.e., we send a matrix $A \in \mathcal{A}_j$ into

$$I \otimes I \otimes \cdots \otimes I \otimes A \otimes I \otimes I \cdots .$$

Our embedding of the algebra of $2^n \times 2^n$ matrices now can be viewed as the tensor product $\mathcal{A}_1 \otimes \mathcal{A}_2 \otimes \cdots \otimes \mathcal{A}_n \otimes I \otimes I \otimes \cdots$ because $M_{2^n}(\mathbb{C})$ is isomorphic to the tensor product of n "copies" of the algebra of 2×2 matrices.

Call $\pi(\mathcal{A}_j)$ the embedding we have made above of $\mathcal{A}_j$ into the jth position of the infinite tensor product. Then the C^*-algebra generated by $\pi(\mathcal{A}_1), \pi(\mathcal{A}_2), \ldots$ is again the CAR algebra. In fact the CAR algebra $\mathfrak{A}$ is sometimes written as

$\mathfrak{A} = \mathcal{A}_1 \otimes \mathcal{A}_2 \otimes \mathcal{A}_3 \otimes \cdots$ but remember that this is nothing more than another way of viewing the norm closure of the sequence of embeddings

$$I \to M_2(\mathbb{C}) \to M_4(\mathbb{C}) \to M_8(\mathbb{C}) \to \cdots$$

that we have described above.

Theorem 1.1. *The CAR algebra $\mathfrak{A}$ is a simple C^*-algebra, i.e., there are no nontrivial norm-closed ideals in $\mathfrak{A}$.*

Proof. Suppose $\mathcal{J}$ is a proper ideal in $\mathfrak{A}$. Then the algebra $\mathfrak{B} = \mathfrak{A}/\mathcal{J}$ becomes a C^*-algebra equipped with the norm $\|\bar{x}\| = \inf_{y \in \mathcal{J}} \|x + y\|$. Let $\phi(x) = \bar{x}$ be the canonical mapping from $\mathfrak{A}$ to $\mathfrak{B}$, then since $\mathfrak{A}_n = M_{2^n}(\mathbb{C})$ is simple for all $n \in \mathbb{N}$, either $\phi(\mathfrak{A}_n)$ is 0 or contains the identity. But $I_{\mathfrak{B}} \in \phi(\mathfrak{A}_n)/(\phi(\mathfrak{A}_n) \cap \mathcal{J})$ so $\phi(\mathfrak{A}_n) \cap \mathcal{J} = \{0\}$, and therefore $\phi : \bigcup_{n=1}^{\infty} \mathfrak{A}_n \to \mathfrak{B}$ is an isometric isomorphism. By continuity so is $\phi : \mathfrak{A} \to \mathfrak{B}$. Hence $\ker(\phi) = \{0\}$ and $\mathfrak{A}$ is simple. □

We conclude our discussion of the CAR algebra by providing yet another description of the algebra which shows how it earns its name. Using this description we then show how the CAR algebra acts as an algebra of bounded operators on the anti-symmetric Fock space $\mathcal{F} = \mathcal{F}(\mathcal{H})$ associated with an underlying separable Hilbert space $\mathcal{H}$.

So let $\mathcal{H}$ be a separable Hilbert space. For each non-negative integer we associate a related Hilbert space $\mathcal{H}_n$. The $\mathcal{H}_n$s are mutually orthogonal. For $n = 1$ the space $\mathcal{H}_1$ is simply $\mathcal{H}$ itself. Now suppose $n = 2$, then H_2 is the Hilbert space spanned by the set of all "wedge" products $f \wedge g$ where f and g are in $\mathcal{H}$. The wedge products satisfy the following identities:

$$c(f \wedge g) = cf \wedge g = f \wedge cg \text{ for } f, g \in \mathcal{H}, c \in \mathbb{C}$$

$$f \wedge g = -g \wedge f$$

$$\langle f \wedge g, h \wedge k \rangle = (f, h)(g, k) - (f, k)(g, h)$$

From the second identity it follows that $f \wedge f = 0$. Using all of the identities it follows that an orthonormal basis for $\mathcal{H}_2$ consists of all wedge vectors $f_i \wedge f_j, i < j$, where $\{f_i : i \in \mathbb{N}\}$ is an orthonormal basis for $\mathcal{H}$.

Similarly one can construct $\mathcal{H}_n$ as a Hilbert space spanned by linear combinations of n-fold wedge products and having an orthonormal basis consisting of all vectors of the form $f_{i_1} \wedge f_{i_2} \wedge \cdots \wedge f_{i_n}$ where $i_1 < i_2 < \cdots < i_n$. Given $2n$ arbitrary vectors $g_1, g_2, \ldots, g_n$ and $h_1, h_2, \ldots, h_n$ in $\mathcal{H}$ the inner product on $\mathcal{H}_n$ satisfies $\langle g_1 \wedge \cdots \wedge g_n, h_1 \wedge \cdots \wedge h_n \rangle = \det(\langle g_i, h_j \rangle)$, the determinant of the $n \times n$ matrix $(\langle g_i, h_j \rangle)$.

The Hilbert space $\mathcal{H}_0$ is a one-dimensional Hilbert space, i.e. it is isomorphic to the complex numbers, and consists of all scalar multiples of Ω, the so-called "vacuum" vector. Ω is a unit vector.

Then *Fock space* $\mathcal{F} = \mathcal{F}(\mathcal{H})$ consists of all vectors $c\Omega \oplus F_1 \oplus F_2 \oplus \cdots$ where $c \in \mathbb{C}$, $F_n \in \mathcal{H}_n$ and $\sum_{j=1}^{\infty} \|F_n\|^2 < \infty$.

How does the CAR algebra act on Fock space? First let's define the so-called creation and annihilation operators $a(f)^*$ and $a(f)$, for $f \in \mathcal{H}$, acting on $\mathcal{F}(\mathcal{H})$ and then see how they in turn generate the CAR algebra.

Definition 1.2. *The creation and annihilation operators $a(f)^*$ and $a(f)$. For $f \in \mathcal{H}$ and for $f_{i_1} \wedge f_{i_2} \wedge \cdots \wedge f_{i_n} \in \mathcal{H}_n$ the vector $a(f)^* f_{i_1} \wedge f_{i_2} \wedge \cdots \wedge f_{i_n}$ is in $\mathcal{H}_{n+1}$ and is given by $a(f)^* f_{i_1} \wedge f_{i_2} \wedge \cdots \wedge f_{i_n} = f \wedge f_{i_1} \wedge f_{i_2} \wedge \cdots \wedge f_{i_n}$. The operator $a(f)^*$ is extended to all of $\mathcal{H}_n$ by linearity. Define $a(f)^*\Omega$ to be f itself. The adjoint is $a(f)$, the so-called annihilation operator, and is given by $a(f)\Omega = 0$ and $a(f) f_{i_1} \wedge f_{i_2} \wedge \cdots \wedge f_{i_n} = \sum_{j=1}^{n} (-1)^{j-1} (f, f_{i_j}) f_{i_1} \wedge \cdots \wedge f_{i_{j-1}} \wedge f_{i_{j+1}} \wedge \cdots \wedge f_{i_n}$ in $\mathcal{H}_{n-1}$.*

Using the definitions above a little work will establish the following identities, the *Canonical Anticommutation Relations*:

$$\{a(f), a(g)\}_+ = a(f)a(g) + a(g)a(f) = 0$$

$$\{a(f)^*, a(g)\}_+ = a(f)^*a(g) + a(g)a(f)^* = \langle f, g \rangle I$$

for all $f, g \in \mathcal{H}$, where I is the identity operator in $\mathcal{B}(\mathcal{F})$.

Using the anticommutation relations above we can manufacture a sequence $\mathcal{A}_j$, $j \in \mathbb{N}$ of 2×2 matrix algebras which pairwise commute and which generate the same algebra in the norm closure as the creation and annihilation operators do, [Po3]. This will allow us to identify our new algebra with the algebra we called the CAR algebra above. First define the following unitary operators. Before looking at the definition it would be helpful to show, using the anticommutation relations above, that for any unit vector f in $\mathcal{H}$ the operator $V = I - 2a(f)^*a(f)$ is a unitary operator, that $V^2 = I$ and that for any vector $g \in \mathcal{H}$ orthogonal to f, V commutes with $a(g)$ and $a(g)^*$.

Definition 1.3 (see [Po3]). *Fix an orthonormal basis $\{f_i : i \in \mathbb{N}\}$ of $\mathcal{H}$. Set V_0 to be the identity operator in $\mathcal{B}(\mathcal{F})$ For any integer $k > 2$, define V_{k-1} to be the product $\Pi_{j=1}^{k-1}(I - 2a(f_j)^*a(f_j))$.*

Theorem 1.4. *For each positive integer k the operator $V_{k-1}a(f_k)$ generates a C^*-algebra $\mathcal{A}_k$ isomorphic to the algebra of 2×2 matrices over $\mathbb{C}$. The algebras $\mathcal{A}_k$ are mutually commuting, i.e., every operator in $\mathcal{A}_k$ commutes with every operator in A_j for $j \neq k$.*

Proof. The remark preceding the definition shows that the hermitian unitary operator V_{k-1} commutes with $a(f_k)$. Therefore the algebra generated by $V_{k-1}a(f_k)$ is 4-dimensional, spanned by $V_{k-1}a(f)$, its adjoint $V_{k-1}a(f)^*$, and

$$a(f_k)^*a(f_k) = (V_{k-1}a(f_k))^* V_{k-1}a(f_k),$$

$$a(f_k)a(f_k)^* = V_{k-1}a(f_k)(V_{k-1}a(f_k))^*.$$

These 4 operators act exactly like (i.e., have the same algebraic properties as) the 4 matrices $\begin{pmatrix} 0 & 1 \\ 0 & 0 \end{pmatrix}$, $\begin{pmatrix} 0 & 0 \\ 1 & 0 \end{pmatrix}$, $\begin{pmatrix} 1 & 0 \\ 0 & 0 \end{pmatrix}$, $\begin{pmatrix} 0 & 0 \\ 0 & 1 \end{pmatrix}$, respectively, so that they generate an algebra $\mathcal{A}_k$ isomorphic to $M_2(\mathbb{C})$.

Using the remark preceding the definition one sees that the algebras $\mathcal{A}_k$ and $\mathcal{A}_j$ are mutually commuting algebras for any $j \neq k$.

Using our identification of the CAR algebra as the algebra constructed as the infinite tensor product, in the sense described above, of an infinite sequence of 2×2 matrix algebras $\mathcal{A}_j$ we have established the following result. □

Corollary 1.5. *The norm-closed algebra generated by the creation and annihilation operators $a(f)^*$ and $a(f)$ for $f \in \mathcal{H}$ is isomorphic to the CAR algebra.*

Recall that a sequence of operators $\{A_n : n \in \mathbb{C}\}$ in $\mathcal{B}(\mathcal{H})$ is said to converge strongly if there is an operator A in $\mathcal{B}(\mathcal{H})$ such that $\|A_n f - Af\|$ tends to 0 for all f in $\mathcal{H}$. It is possible to show (although we don't include the proof) that the following holds for the CAR algebra and its representation as an algebra of operators acting on the Fock space $\mathcal{F}(\mathcal{H})$.

Theorem 1.6. *The strong closure of the CAR algebra in $\mathcal{B}(\mathcal{H})$ coincides with $\mathcal{B}(\mathcal{H})$ itself.*

Note that although the CAR algebra contains no nonzero compact operators the theorem above shows that every compact operator is contained within its strong closure!

2. Unbounded symmetric linear transformations on Hilbert space

In the late 1920s M. H. Stone established a remarkable connection between strongly continuous one-parameter groups of unitary operators on a Hilbert space $\mathcal{H}$ and self-adjoint operators on $\mathcal{H}$. Stone's result proves that the generator of a one-parameter group of unitaries is a self-adjoint operator. Conversely, given a self-adjoint operator A on a Hilbert space $\mathcal{H}$, A is the generator of a one-parameter group of unitary operators.

One of our goals is to define symmetric operators on $\mathcal{H}$. The self-adjoint operators are a special case of these. We introduce von Neumann's elegant criterion for determining when a symmetric operator is self-adjoint. More generally von Neumann's results tell us when a symmetric operator generates a strongly continuous one-parameter semigroup of isometries on a Hilbert space $\mathcal{H}$. Later we shall use this result to construct the most basic of the E_0-semigroups on $\mathcal{B}(\mathcal{H})$, which are known as the CAR flows.

Definition 2.1. *Let $\mathcal{D}$ be a densely-defined linear subspace of a Hilbert space $\mathcal{H}$. We say that a linear operator from $\mathcal{D}$ to $\mathcal{H}$ is a symmetric linear transformation on $\mathcal{D}$ if for all f, g in $\mathcal{D}$,*

$$(Af, g) = (f, Ag).$$

Here are three examples, with more to follow! (Here we can take the term *absolutely continuous* to describe a function which is the antiderivative of a function in L^2, [RSz].)

Example 2.1. Let $\mathcal{H}$ be $L^2(0, 1)$. Consider the dense linear subspace $\mathcal{D}_1$ of $\mathcal{H}$ given by $\mathcal{D}_1 = \{f : f$ is absolutely continuous, $f(0) = f(1) = 0$, and $f' \in \mathcal{H}\}$. Let $A_1 = id/dx$ on its domain $\mathcal{D}(A_1) = \mathcal{D}_1$. Integration by parts shows that A_1 is a symmetric operator on $\mathcal{H}$.

Example 2.2. Same $\mathcal{H}$ and $A_2 = id/dx$ again but this time set $\mathcal{D}(A_2) = \mathcal{D}_2 = \{f : f$ is absolutely continuous, $f(0) = f(1)$, and $f' \in \mathcal{H}\}$. Another exercise in integration by parts establishes the fact that A_2 is symmetric.

Remark. Since $\mathcal{D}_1 \subset \mathcal{D}_2$ and since A_2 coincides with A_1 on $\mathcal{D}_1$ we say that the symmetric operator A_2 is an *extension* of the symmetric operator A_1, cf. Definition 2.7.

Example 2.3. Same $\mathcal{H}$. $\mathcal{D}_3 = \{f : f$ is absolutely continuous and $f' \in \mathcal{H}\}$.

It's pretty clear that A_1 is an *unbounded* operator. In fact if we take $f_n(x) = x^n$ then $\|A_1 f_n\|/\|f_n\| = \frac{n/\sqrt{2n-2}}{1/\sqrt{2n+1}} = o(n^2)$. Of course the derivative operator id/dx is not defined everywhere. Is there any possibility that an unbounded symmetric operator could have domain $\mathcal{D} = \mathcal{H}$? The answer is no.

Theorem 2.2 (Hellinger-Toeplitz, [RSz]). *Suppose A is a linear symmetric operator on $\mathcal{H}$ with domain $\mathcal{D}$ equal to all of $\mathcal{H}$. Then A must be a bounded operator, i.e., there is an $r > 0$ such that $\|Af\| \leq r\|f\|$ for all $f \in \mathcal{H}$.*

Dealing with domains of unbounded operators can be tricky but the following rules for combining such operators are rather commonsensical.

1. If $c \in \mathbb{C}$ then cA has the same domain as A (even if $c = 0$!) and $(cA)(f) = c(Af)$.
2. If T_1 and T_2 are unbounded operators then $T_1 + T_2$ has domain $\mathcal{D}(T_1) \cap \mathcal{D}(T_2)$ (which may not be dense in $\mathcal{H}$, in fact it may be $\{0\}$!).
3. If T_1 and T_2 are unbounded operators then $\mathcal{D}(T_1T_2) = \{f \in \mathcal{D}(T_2) : T_2 f \in \mathcal{D}(T_1)\}$ and $(T_1T_2)(f) = T_1(T_2 f)$. It's possible for $\mathcal{D}(T_1)$ to be dense and yet $\mathcal{D}(T_2) = \{0\}$!

Inverses. Suppose A is a one-one transformation then $\mathcal{D}(A^{-1}) = Ran(A)$ and of course $A^{-1}(Af) = f$. We remark that A^{-1} might be bounded in norm. Consider the example where $\{f_n : n \in \mathbb{N}\}$ is an orthonormal basis for $\mathcal{H}$, $\mathcal{D}(A)$ consists of all vectors $\sum_{n=1}^{\infty} c_n f_n$ in $\mathcal{H}$ for which $\sum_{n=1}^{\infty} n^2 c_n f_n$ converges. Set $Af_n = n^2 f_n$. It's pretty clear that A is a symmetric linear operator. In addition, just looking at the vectors in the basis shows us that A is unbounded. A has an inverse which is defined on the f_ns by $A^{-1} f_n = \frac{1}{n^2} f_n$. Let $h \in \mathcal{H}$, then there are scalars $c_n \in \mathbb{C}$ with $(c_n) \in \ell^2(\mathbb{N})$ such that $h = \sum_{n=1}^{\infty} c_n f_n$ and $h = Af$ where $f = \sum_{n=1}^{\infty} \frac{c_n}{n^2} f_n \in \mathcal{D}(A)$. Hence $Ran(A) = A\mathcal{D}(A) = \mathcal{H}$.

Adjoints. We know from experience (or from the first lecture) that if $T \in M_n(\mathbb{C})$, the matrix S that satisfies the equation $(Tf, g) = (f, Sg)$ is $S = T^* = \bar{T}^t$, the conjugate transpose of T. If T is a bounded operator on $\mathcal{H}$ then viewing T as an infinite matrix as we did in the first lecture we observe once again that T^* is the conjugate transpose of the infinite matrix corresponding to T. Defining adjoints for unbounded operators can be tricky, but their study makes for some very interesting mathematics. So suppose T is an unbounded operator with dense domain $\mathcal{D}(T) \subset \mathcal{H}$.

Definition 2.3. *The adjoint T^* of T. Let T be an unbounded linear transformation on $\mathcal{H}$ with dense domain $\mathcal{D}(T)$. Then $\mathcal{D}(T^*)$ consists of all vectors $g \in \mathcal{H}$ for which the map $f \in \mathcal{D}(T) \to (Tf, g)$ is continuous. By the Hahn-Banach Theorem we can extend the functional above to a continuous functional on all of $\mathcal{H}$ and therefore there is a vector g^* in $\mathcal{H}$ for which $(Tf, g) = (f, g^*)$ for all $f \in \mathcal{D}(T)$. Then we define T^*g to be g^*.*

Let's look at some adjoints by returning to our examples (see [R]).

Example 2.4. $A_1 = id/dx$ on its domain $\mathcal{D}(A_1)$ consisting of all absolutely continuous functions f on the unit interval with endpoint conditions $f(0) = 0 = f(1)$.

Suppose g is any absolutely continuous function on the unit interval, then for $f \in \mathcal{D}(A_1)$,

$$
\begin{aligned}
(A_1 f, g) = (if', g) &= \int_0^1 if'(x)\bar{g}(x)dx \\
&= (if(1)\bar{g}(1) - if(0)\bar{g}(0)) - \int_0^1 if(x)\bar{g}'(x)dx \\
&= \int_0^1 f(x)\overline{ig'(x)}dx \\
&= (f, ig').
\end{aligned}
$$

This shows us that $\{g \in L^2(0,1) : g$ is absolutely continuous and $g' \in L^2(0,1)\}$ is a subset of $\mathcal{D}(A_1^*)$ and that $A_1^* = id/dx$ on this set. We'll do a little more work in the next paragraph to establish that the set above actually coincides with $\mathcal{D}(A_1^*)$. In the meantime note that A_1^* is not symmetric, however, since for any f, g in the set above, $(A_1^* f, g) - (f, A_1^* g) = if(1)\bar{g}(1) - if(0)\bar{g}(0)$, which is not necessarily 0.

Let's show that A_1^* is the operator id/dx with domain given in the previous paragraph. Suppose $g \in \mathcal{D}(A_1^*)$ and $h = A_1^* g$. Let $k(x) = \int_0^x h(x)dx$. Then for $f \in \mathcal{D}(A_1)$ we compute $-\int_0^1 f'(x)\overline{ig(x)}dx = \int_0^1 if'(x)\bar{g}(x)dx = (A_1 f, g) = (f, A_1^* g) = (f, h)$ and $(f, h) = \int_0^1 f(x)\bar{h}(x)dx = f(1)\bar{k}(1) - f(0)\bar{k}(0) - \int f'(x)\bar{k}(x)dx$. But f has zero boundary conditions so $(f, h) = -\int f'(x)\bar{k}(x)dx$. Therefore $\int_0^1 f'(x)(\overline{ig(x)} - \overline{k(x)})dx = 0$ for all f in $\mathcal{D}(A_1)$. But $Ran(A_1)$ can be shown to contain all functions u such that $\int_0^1 u(x)dx = 0$. On the other hand, any function y in L^2 can be written as the sum of a constant function with a function u whose integral is 0 over $[0,1]$. Therefore we conclude that $ig(x) - k(x)$ is a constant function. Hence $-ik(x) = g(x) - c$ for some constant function c. Hence g is absolutely continuous and $ig' = k' = h$, i.e., $A_1^* g = ig'$.

We have shown the following.

Theorem 2.4. $A_1^* = A_3$.

Theorem 2.5. *Suppose $g_n, n \in \mathbb{N}$ is a Cauchy sequence of vectors in the domain $\mathcal{D}(T^*)$ of the adjoint of a linear operator T, and suppose $T^* g_n$ also converges to a vector g^*. Then if $g_n \to g$, $g \in \mathcal{D}(T^*)$ and $T^* g = g^*$.*

Proof. We have $(f, g^*) = \lim_{n\to\infty}(f, T^* g_n) = \lim_{n\to\infty}(Tf, g_n) = (Tf, g)$. This shows that the mapping $f \to (Tf, g)$ is continuous. Hence $g \in \mathcal{D}(T^*)$ and since $(f, g^*) = (Tf, g)$, $g^* = T^* g$. □

Definition 2.6. *An unbounded operator T with domain $\mathcal{D}(T) \subset \mathcal{H}$ is said to be closed if whenever $f_n \in \mathcal{D}(T)$, $Tf_n \in \mathcal{H}$ are both convergent sequences with limits f, h, respectively, then $f \in \mathcal{D}(T)$ and $Tf = h$.*

Corollary 2.7. *The adjoint of a densely-defined linear operator is closed.*

Definition 2.8. *Let S and T be unbounded operators. If $\mathcal{D}(S) \subset \mathcal{D}(T)$ and $Tf = Sf$ for all $f \in \mathcal{D}(S)$ we say that T is an extension of S. We use the notation $S \subset T$ to denote this relationship.*

From the definition of a symmetric operator S we always have $S \subset S^*$. In fact we also have the following.

Theorem 2.9. *Suppose T is a closed densely-defined linear operator on a Hilbert space $\mathcal{H}$. Then*

1. *T^* is densely-defined and*
2. *$T = T^{**}$.*

Remark. We will closely follow the proof in Riesz-Sz.-Nagy, [RSz], Chapter VIII, Section 117, which relies on the notion of the graph $\mathcal{G}_T$ of a linear operator T on $\mathcal{H}$. To define $\mathcal{G}_T$ we must first introduce the set $\bar{\mathcal{H}}$ of all pairs $\{f, g\}$ of vectors f and g in $\mathcal{H}$. We turn $\bar{\mathcal{H}}$ into a linear space by setting

$$c\{f, g\} = \{cf, cg\},$$
$$\{f_1, g_1\} + \{f_2, g_2\} = \{f_1 + f_2, g_1 + g_2\},$$

and we turn $\bar{\mathcal{H}}$ into a Hilbert space by setting

$$(\{f_1, g_1\}, \{f_2, g_2\}) = (f_1, f_2) + (g_1, g_2).$$

Then the graph $\mathcal{G}_T$ of a linear transformation T is the subspace of $\bar{\mathcal{H}}$ consisting of all vectors $\{f, Tf\}$ where $f \in \mathcal{D}(T)$. It is not difficult to show that $\mathcal{G}_T$ is closed if and only if T is a closed operator.

Proof of Theorem 2.8. Let V be the linear mapping on $\bar{\mathcal{H}}$ given by $V\{f, g\} = \{g, -f\}$. It's clear that V is a unitary operator (note that $V^2 = -I$ where I is the identity on $\bar{\mathcal{H}}$). Now suppose $g \in \mathcal{D}(T^*)$ and $g^* = T^*g$ then for all $f \in \mathcal{D}(T)$,

$$\begin{aligned}
(Tf, g) = (f, g^*) &\iff (Tf, g) - (f, g^*) = 0 \\
&\iff (\{Tf, -f\}, \{g, g^*\}) = 0 \\
&\iff (V\{f, Tf\}, \{g, g^*\}) = 0 \\
&\iff \{g, g^*\} \in \bar{\mathcal{H}} \ominus [V\mathcal{G}_T], \text{ where [ ] means closure} \\
&\iff \{g, g^*\} \in \bar{\mathcal{H}} \ominus V[\mathcal{G}_T].
\end{aligned}$$

Since we're assuming that T is closed the last line above becomes

$$\mathcal{G}_{T^*} = \bar{\mathcal{H}} \ominus V\mathcal{G}_T.$$

If $\mathcal{D}(T^*)$ is not densely-defined then there is an $h \in \mathcal{H}$ such that h is orthogonal to $\mathcal{D}(T^*)$, which implies that $\{T^*g, -g\} \perp \{0, h\}$ for all $g \in \mathcal{D}(T^*)$, i.e., $\{0, h\} \perp V\mathcal{G}_T^*$. But $\mathcal{G}_{T^*} = \bar{\mathcal{H}} \ominus V\mathcal{G}_T$, and so

$$\begin{aligned} V\mathcal{G}_{T^*} &= V\bar{\mathcal{H}} \ominus V^2\mathcal{G}_T \\ &= \bar{\mathcal{H}} \ominus \mathcal{G}_T. \end{aligned}$$

Thus $\{0, h\} \in \mathcal{G}_T$ and so $h = T(0) = 0$.

Let's finish by showing that T coincides with T^{**} under the assumption that T is a closed operator. We have

$$\begin{aligned} \mathcal{G}_{T^{**}} &= \bar{\mathcal{H}} \ominus V\mathcal{G}_{T^*} \\ &= \bar{\mathcal{H}} \ominus V(\bar{\mathcal{H}} \ominus V\mathcal{G}_T) \\ &= \bar{\mathcal{H}} \ominus (\bar{\mathcal{H}} \ominus \mathcal{G}_T) \\ &= \mathcal{G}_T, \end{aligned}$$

and we're done. □

Theorem 2.10. *If S is an unbounded symmetric operator then S^{**} is also. And $S \subset S^{**}$.*

Proof. Let f be in $\mathcal{D}(S)$ then the mapping $g \to (f, S^*g) = (Sf, g)$ is continuous in g for g in $\mathcal{D}(S^*)$ and so $f \in \mathcal{D}((S^*)^*) = \mathcal{D}(S^{**})$. Hence $S \subset S^{**}$. On the other hand, since $S \subset S^*$ upon taking adjoints we get $S^* \supset S^{**}$, hence $S^{**} \subset S^* = (S^*)^{**}$, where we are using the fact that if T is a closed operator then $T = T^{**}$. Therefore $S \subset S^{**}$ and S^{**} is symmetric. □

Example 2.5. We shall show that the linear operator A_2 is actually self-adjoint, i.e., $A_2 = A_2^*$. Note that the boundary conditions on functions in $\mathcal{D}(A_2)$ are looser than those on $\mathcal{D}(A_1)$ but more restrictive than those imposed on $\mathcal{D}(A_3)$. We might express this by saying that the boundary conditions on $\mathcal{A}_1$ are too severe for A_1 to be self-adjoint and those on A_3 are too free for A_3 to be self-adjoint. The boundary conditions on A_2 are just right!

Theorem 2.11. *A_2 is a self-adjoint operator, i.e., $A_2 = A_2^*$.*

Proof. We follow much the same argument that we used above to show that $A_1^* = A_3$. First observe that A_2 is symmetric. In other words, $(A_2f, g) = (f, A_2g)$, and this is easily established using integration by parts. Next suppose $g \in \mathcal{D}(A_2^*)$ and $A_2^*g = h$. As above define $k(x)$ to be $\int_0^x h(t)dt$. Then as above we calculate $-\int_0^1 f'(x)\overline{ig(x)}dx = \int_0^1 if'(x)\bar{g}(x)dx = (A_2f, g) = (f, A_2^*g) = (f, h)$ and $(f, h) = \int_0^1 f(x)\bar{h}(x)dx = f(1)\bar{k}(1) - f(0)\bar{k}(0) - \int f'(x)\bar{k}(x)dx$. We have $k(0) = 0$ from the definition of k. Choosing f to be any nonzero constant function shows that $k(1) = 0$, so we have, again, $(f, h) = \int_0^1 f'(x)\overline{ig(x)}dx = \int_0^1 f'(x)\bar{k}(x)dx$. Thus, as above, $ig = k + c$, so g is absolutely continuous. Also $g(0) = g(1)$ since $k(0) = k(1) = 0$, so g is actually in $\mathcal{D}(A_2)$. Therefore A_2 is self-adjoint. □

We end by adding one more example to our list of symmetric operators.

Example 2.6. Let A_4 be the operator id/dx on its domain in $L^2(0,\infty)$ consisting of all absolutely continuous functions f such that $f(0)=0$ and such that f' is back in L^2.

3. Deficiency indices and extensions of symmetric operators

Consider the linear fractional transformation $\phi(z)$ that takes $z\in\mathbb{C}, z\neq -i$ to $\frac{z-i}{z+i}$. We recall from the theory of complex variables that linear fractional transformations $z\to\frac{az+b}{cz+d}$ take circles to lines or circles and lines to circles or lines. In our example the real axis (plus ∞) is mapped onto the unit circle in the complex plane.

What happens to a diagonal self-adjoint matrix $A\in M_n(\mathbb{C})$ under the mapping $A\to\phi(A)=\frac{A-iI}{A+iI}$? A little work shows that $\operatorname{diag}(x_1,x_2,\ldots,x_n)$ maps to $\operatorname{diag}(\phi(x_1),\phi(x_2),\ldots,\phi(x_n))$, a unitary diagonal matrix. It was von Neumann's idea to consider analogues of this mapping for unbounded symmetric operators, from which one can derive a very elegant necessary and sufficient condition for closed symmetric operator to be self-adjoint.

For the rest of this section we assume that A is a closed symmetric operator on a Hilbert space $\mathcal{H}$ with dense domain $\mathcal{D}(A)$.

Theorem 3.1. *$(A\pm iI)\mathcal{D}(A)$ are closed subspaces of $\mathcal{H}$.*

Proof. Note $\|(A\pm iI)f\|^2=\|Af\|^2+\|f\|^2\geq\|f\|^2$ for all $f\in\mathcal{D}(A)$. Suppose then that $\{f_n:n\in\mathbb{N}\}$ is a sequence in $\mathcal{D}(A)$ such that the sequence $(A+iI)f_n$ converges (to h). We'd like to show that f is in $\mathcal{D}(A)$ and that $Af=h$. The calculation above shows that $\{f_n:n\in\mathbb{N}\}$ is a Cauchy sequence. Let f be its limit.

Since $(A+iI)f_n\to h$ and since $f_n\to f$ it follows that $Af_n\to h-if$. Therefore $f\in\mathcal{D}(A)$ because A is a closed operator, by assumption. Hence $(A+iI)f=h$ and $(A+iI)\mathcal{D}(A)$ is closed subspace of $\mathcal{H}$. Similarly one can show that $(A-iI)\mathcal{D}(A)$ is closed in $\mathcal{H}$. □

Definition 3.2. *The Cayley transform. Let V be a linear mapping with domain $(A+iI)\mathcal{D}(A)$ and range $(A-iI)\mathcal{D}(A)$ given by $V(A+iI)f=(A-iI)f$.*

From the proof of the preceding theorem V is an *isometry* from the closed subspace $(A+iI)\mathcal{D}(A)$ to $(A-iI)\mathcal{D}(A)$, i.e., V preserves norm.

Example 3.1. Consider the symmetric operator A_2 from Example 2.2. Let's find the orthogonal complement of the two closed subspaces $(A_2+iI)\mathcal{D}(A_2)$ and $(A_2-iI)\mathcal{D}(A_2)$. If $g\in(A_2+iI)\mathcal{D}(A_2)^{\perp}$ then for all $f\in\mathcal{D}(A_2)$, $(A_2f+if,g)=(if'+if,g)=0$, so $(f'+f,g)=0$ for all $f\in\mathcal{D}(A_2)$. For any h in $L^2(0,1)$, however $f(x)=c\cdot\exp(-x)+\exp(-x)\int_0^x\exp(t)h(t)dt$ are the solutions to the differential equation $f'+f=h$ and c can be chosen so that $f(0)=f(1)$. Similarly for the differential equation $f'-f=h$. Hence both $(A_2+iI)\mathcal{D}(A_2)$ and $(A_2-iI)\mathcal{D}(A_2)$ coincide with all of $\mathcal{H}=L^2(0,1)$. The Cayley transform V is in this case a unitary operator from $\mathcal{H}$ onto $\mathcal{H}$, taking $f'+f$ to $f'-f$, for all $f\in D(A_2)$.

Example 3.2. Let's find the orthogonal complement to $(A_1 + iI)\mathcal{D}(A_1)$. Looking back at our calculations in Example 1 we see that we're looking for all $g \in \mathcal{H}$ which satisfy $(f' + f, g) = 0$. Note that this implies that the mapping $f \to (f', g)$ is continuous for $f \in \mathcal{D}(A_1)$ and therefore $g \in \mathcal{D}(A_3)$ because A_1^* is A_3. This tells us that g has a derivative, so

$$0 = (f' + f, g) = (f, g - g'),$$

and since $\mathcal{D}(A_1)$ is dense in L^2 we conclude that $g = g'$, i.e., $g(x) = ce^x$. Thus the complement of the range of $A_1 + iI$ is spanned by the single function e^x. Similarly the complement of $Ran(A_1 - iI)$ is spanned by e^{-x}.

Consider the following questions.

1. Given a symmetric linear operator A on $\mathcal{H}$, when is A self-adjoint?
2. Suppose A is symmetric but not self-adjoint. Are there any self-adjoint extensions T of A? If not, why not?
3. Suppose A is symmetric and has self-adjoint extensions. Can we classify all of the self-adjoint extensions?

The theorem below (due to von Neumann) will answer all of these questions. First we require the following definition.

Definition 3.3. *Deficiency spaces and deficiency indices. Let A be a closed unbounded symmetric operator with dense domain $\mathcal{D}(A)$ in a Hilbert space $\mathcal{H}$. The deficiency spaces $\mathcal{D}_+$ and $\mathcal{D}_-$ of A are the orthogonal complements of the closed subspaces $(A + iI)\mathcal{D}(A)$ and $(A - iI)\mathcal{D}(A)$, respectively. The deficiency indices (n_+, n_-) are the dimensions of the orthogonal complements $\mathcal{D}_+$, $\mathcal{D}_-$ above. Hence $0 \leq n_+ \leq \infty$ and $0 \leq n_- \leq \infty$.*

Theorem 3.4. *Let A be a closed unbounded symmetric operator with deficiency indices (n_+, n_-). Then*

1. *A is self-adjoint if and only if $n_+ = 0 = n_-$.*
2. *A is maximal symmetric (i.e., A has no proper symmetric extensions) if and only if $n_+ = 0$ or $n_- = 0$ (or both).*
3. *If $n_+ = n_-$ then the self-adjoint extensions T of A are in one-one correspondence with the set of all isometries from $\mathcal{D}_+$ to $\mathcal{D}_-$. The corresponding isometric extensions $\tilde{V}$ of V are the Cayley transforms of the self-adjoint extensions.*
4. *If $0 < n_+ < n_-$ then the maximal symmetric extensions of A correspond to the set of all isometries from $\mathcal{D}_+$ to subspaces of $\mathcal{D}_-$.*
5. *If $0 < n_- < n_+$ then the maximal symmetric extensions of A correspond to the isometric mappings from subspaces of dimension n_- of $\mathcal{D}_+$ onto $\mathcal{D}_-$.*

Example 3.3. For our earlier example A_1, We have that $n_+ = 1 = n_-$. Here $\mathcal{D}_+$ is spanned by e^x and $\mathcal{D}_-$ is spanned by e^{-x}. Let's calculate all of the self-adjoint extensions of A_1. Since $\mathcal{D}_+$ and $\mathcal{D}_-$ are one-dimensional this will be pretty easy. First note that since $A_1 \subset A_3$ all of the extensions will act like id/dx on their domains, and their domains will be subsets of $\mathcal{D}(A_3)$. Next note that $\phi_+(x) = \sqrt{\frac{2}{e^2-1}}e^x$ and

$\phi_- = e\sqrt{\frac{2}{e^2-1}}e^{-x}$ are unit vectors. The isometries from $\mathcal{D}_+$ to $\mathcal{D}_-$ are therefore of the form $\phi_+ \to -\gamma\phi_-$ for any γ on the unit circle in the complex plane (the minus sign is merely a convenience). Then if $\tilde{V} = V_\gamma$ is the corresponding isometry extending V, and $A = A(\gamma)$ is the corresponding self-adjoint extension of A_1,

$$\tilde{V} = (A - iI)(A + iI)^{-1}.$$

This suggests that

$$A = i(I + \tilde{V})(I - \tilde{V})^{-1}.$$

Now $(I - \tilde{V})\phi_+ = \phi_+ + \gamma\phi_-$ so $\phi_+ + \gamma\phi_-$ is in the domain of A and

$$\begin{aligned} A(\phi_+ + \gamma\phi_-) &= i(d/dx)(\phi_+ + \gamma\phi_-) \\ &= i(\phi_+ - \gamma\phi_-) \end{aligned}$$

Call $\psi = \phi_+ + \gamma\phi_-$ and note that $\psi(0) = i\left(\sqrt{\frac{2}{e^2-1}} + \gamma\sqrt{\frac{2}{e^2-1}}e\right)$, and $\psi(1) = i\left(\sqrt{\frac{2}{e^2-1}}e + \gamma\sqrt{\frac{2}{e^2-1}}\right)$. Then

$$\frac{\psi(1)}{\psi(0)} = \frac{1 + \gamma e}{e + \gamma} = \alpha,$$

and $|\alpha| = 1$, so we conclude that the domain $\mathcal{D}(A(\gamma))$ of $A(\gamma) = A$ consists of all absolutely continuous functions $f \in L^2$ such that $f(1) = \alpha f(0)$ where $\alpha = \frac{1-\gamma e}{e-\gamma}$.

Note that our self-adjoint operator A_2 from Example 2.2 is one of these. In fact its domain consists of all absolutely continuous functions f in L^2 with $f(0) = f(1)$. Then $\alpha = 1$, so $\gamma = -1$, and $\psi = \phi_+(x) - \phi_-(x)$.

Example 3.4. A_4 is the operator id/dx on its domain in $L^2(0, \infty)$ consisting of all absolutely continuous functions f such that $f(0) = 0$ and such that f' is back in L^2.

1. Show that A_4 is symmetric.
2. Show that A_4^* consists of all absolutely continuous functions f on L^2 with $f(0)$ arbitrary and f' in L^2.
3. Show that $\mathcal{D}_+ = \{0\}$ and $\mathcal{D}_-$ consists of scalar multiples of e^{-x}. Conclude that A_4 is maximal symmetric.
4. Let V be the Cayley transform. Show that V is unitarily equivalent to the unilateral shift S defined on $\ell^2(\mathbb{N})$ by $Se_n = e_{n+1}$ where $\{e_n : n \in \mathbb{N}\}$ is an orthonormal basis for ℓ^2.

In $M_n(\mathbb{C})$ there is no distinction between symmetric and self-adjoint, where by symmetric we mean, of course $A = \bar{A}^t$. They mean the same thing. Suppose A is a (necessarily bounded, necessarily everywhere-defined) self-adjoint matrix. Then we can exponentiate A to form the matrix e^{itA} for each t, and it is not difficult to show that the resulting matrix is a unitary matrix. In fact we obtain a one-parameter family $\{U_t = e^{itA} : t \in \mathbb{R}\}$ of unitary operators satisfying the identities $U_0 = I, U_sU_t = U_{s+t}$

for all $s, t \in \mathbb{R}$. These results can easily be derived by using the spectral theorem to put A into diagonal form: if $A = diag(\lambda_1, \ldots, \lambda_n)$ then $U_t = diag(e^{it\lambda_1}, \ldots, e^{it\lambda_n})$. What happens in infinite dimensions? The same thing, if one assumes that A is a bounded self-adjoint operator: A is said to generate the one-parameter group of (norm) continuous unitary operators $U_t, t \in \mathbb{R}$. On the other hand, in many physical applications we are interested in a one-parameter group of strongly continuous unitary operators, i.e., the mapping $t \to U_t f$ is continuous in t for all f in the Hilbert space. Here as in the bounded case we can try to take the derivative of the unitary group. More specifically we look for all f in $\mathcal{H}$ for which the limit

$$\lim_{t \to 0} \frac{U_t f - f}{t}$$

exists. It turns out that the set of f for which this limit exists is a dense linear submanifold of $\mathcal{H}$. In fact the derivative is iA where A is a self-adjoint operator on $\mathcal{H}$. Here is Stone's famous theorem on unitary groups acting on $\mathcal{H}$. We will have to leave the statement of the theorem a little vague, since a precise definition of the operators e^{itA} in the theorem would lead us into the spectral theory of self-adjoint operators, which is a rather large project in and of itself.

Theorem 3.5 (M. H. Stone, 1929). *Let U_t be a strongly continuous one-parameter group of unitary operators on a Hilbert space $\mathcal{H}$. Then there is a self-adjoint operator A on $\mathcal{H}$ such that $U_t = e^{itA}$.*

We say that A is the generator of the U_t's. Recall that we have just found all of the self-adjoint extensions of the symmetric operator A_1 from Example 2.2. What is the unitary group that each one generates? Suppose A is the self-adjoint extension of A_1 whose domain consists of all absolutely continuous functions f with f' in $L^2(0, 1)$ and satisfying the boundary condition $f(0) = \alpha f(1)$, for $|\alpha| = 1$. If f is an L^2 function supported away from both 0 and 1 then $U_t f$ acts like translation to the right, for small t. What happens when we bump up against the boundary point 1? Then the function feeds back into the left side of the unit interval, except it comes in with a phase shift of α. For example suppose f is the characteristic function $\chi\left(\frac{1}{3}, \frac{2}{3}\right)$ on the interval $[1/3, 2/3]$. Then for $0 \leq t \leq 1/3$, $U_t f$ is the characteristic function on the interval $\left[\frac{1}{3} + t, \frac{2}{3} + t\right]$. Suppose $t = \frac{1}{2}$, then $U_{\frac{1}{2}} f = \chi\left(\frac{5}{6}, 1\right) + \alpha\chi\left(0, \frac{1}{3}\right)$.

To conclude this lecture let's consider Example 2.4 once again. Let $S_t, t \in [0, \infty)$ be the semigroup of strongly continuous unilateral shifts on $L^2(0, 1)$, i.e.,

$$(S_t f)(x) = f(x - t), \text{ for } x \geq t, \text{ and}$$

$$(S_t f)(x) = 0, \text{ for } 0 \leq x < t.$$

Then the semigroup $\{S_t : t \geq 0\}$ is a semigroup of isometries (not unitaries!) for $t > 0$. For $f \in \mathcal{D}(A_4)$ we calculate

$$\lim_{t \to 0} \frac{(S_t f)(x - t) - f(x)}{t} = -f'(x) = (iA_4 f)(x)$$

This suggests that iA_4 is the infinitesimal generator of S_t, and it is. Moreover, one has the following result.

Theorem 3.6. *Let A be a maximal symmetric operator on a Hilbert space $\mathcal{H}$ with deficiency indices $(0, n_-)$, where $n_- > 0$. Then A is the generator of a strongly continuous one-parameter semigroup $\{V_t : t \geq 0\}$ of isometries of the form $V_t = U_t \oplus_{k=1}^{n_-} S_t$, i.e., the Hilbert space splits into a direct sum in such a way that V_t consists of a unitary part along with the direct sum of n_- copies of the unilateral shift above.*

4. Derivations

In the Ising model and other models of physical systems the self-adjoint operators of a C^*-algebra $\mathfrak{A}$ represent the observables of the system. In the Heisenberg approach to dynamical systems the operators evolve in time via a one-parameter group of $*$-automorphisms of the algebra, $\{\alpha_t : t \in \mathbb{R}\}$, i.e., if A is an observable at time $t = 0$ then $\alpha_t(A)$ is the evolution of the observable at time t. It turns out that the generator (derivative) of the one-parameter group α_t, $t \in \mathbb{R}$ is a derivation, which we define below. Usually the derivation δ is an unbounded operator and we have to worry about the domain of definition of the derivation. For the moment we will content ourselves with a discussion of bounded derivations on a C^*-algebra.

Definition 4.1. *Let $\mathfrak{A}$ be a C^*-algebra. A bounded derivation is a bounded operator δ on $\mathfrak{A}$ which satisfies the Leibniz rule*

$$\delta(xy) = \delta(x)y + x\delta(y),$$

for all x, y in $\mathfrak{A}$. The derivation δ is said to be a $*$-derivation, or a symmetric derivation, if $\delta(x^*) = \delta(x)^*$ for all x in $\mathfrak{A}$.

Examples of bounded derivations are easy to come by, at least for noncommutative C^*-algebras.

Example 4.1. If a is any element of a C^*-algebra $\mathfrak{A}$ define $\delta = \delta_a$ on $\mathfrak{A}$ by

$$\delta(b) = [a, b],$$

for $b \in \mathfrak{A}$. If h is a hermitian (i.e., self-adjoint) operator in $\mathfrak{A}$ then δ_{ih} is easily seen to be a $*$-derivation.

Are there other types of bounded derivations? Not in matrix algebras! The following result is attributed to Emmy Noether.

Theorem 4.2. *If $\mathfrak{A} = M_n(\mathbb{C})$, the algebra of $n \times n$ matrices over $\mathbb{C}$ then any derivation on $\mathfrak{A}$ is a commutator.*

Proof. Let δ be a derivation on $\mathfrak{A}$. Let $\{e_{ij} : 1 \leq i, j \leq n\}$ be matrix units, i.e., let e_{ij} represent the matrix with a 1 in the (i, j) position and 0s everywhere else. Then the matrix units satisfy the identities $e_{ij}e_{pq} = \delta_{jp}e_{iq}$ (where δ_{jp} is the Kronecker delta). Let k be the matrix

$$k = \sum_{i=1}^{n} \delta(e_{i1})e_{1i}.$$

Observe that $\delta(1) = \delta(1^2) = \delta(1)1 + 1\delta(1)$ so that $\delta(1) = 0$ and therefore

$$0 = \delta(1) = \sum_{i=1}^{n} \delta(e_{ii})$$

$$\sum_{i=1}^{n} \delta(e_{i1}e_{1i}) = \sum_{i=1}^{n} \delta(e_{i1})e_{1i} + \sum_{i=1}^{n} e_{i1}\delta(e_{1i}) = k + -k.$$

Therefore, for any matrix unit e_{pq} we get

$$\begin{aligned}[k, e_{pq}] &= \sum_{i=1}^{n} \delta(e_{i1})e_{1i}e_{pq} + \sum_{i=1}^{n} e_{pq}e_{i1}\delta(e_{1i}) \\ &= \delta(e_{p1})e_{1q} + e_{p1}\delta(e_{1q}) \\ &= \delta(e_{p1}e_{1q}) = \delta(e_{pq}),\end{aligned}$$

so that $\delta(a) = [k, a]$ for all $a \in \mathfrak{A}$, by linearity. □

Question. Are all bounded derivations given by commutators on a C^*-algebra $\mathfrak{A}$?

The answer is yes in many instances.

Theorem 4.3 (K, S2, S3). *Yes, if $\mathfrak{A}$ is a von Neumann algebra.*

Theorem 4.4 (S1, S3). *Yes, on simple C^*-algebras with identity.*

Theorem 4.5 (SW, S3). *Yes, on commutative C^*-algebras, in the sense that there are no nontrivial bounded derivations on commutative C^*-algebras.*

Next we turn to unbounded derivations. As mentioned above we have to exercise care with unbounded derivations because they are not defined everywhere. In fact, [Sakai, Tohoku M. J. vol. 12 (1960)] if δ is an everywhere-defined derivation on a C^*-algebra $\mathfrak{A}$ then δ is automatically a bounded derivation.

Definition 4.6. *Let $\mathcal{D}(\delta)$ be a dense (in a suitable topology) $*$-subalgebra of a C^*-algebra $\mathfrak{A}$ then a derivation is a linear operator from $\mathcal{D}(\delta)$ to $\mathfrak{A}$ which satisfies the Leibniz identity for all products xy, where $x, y \in \mathcal{D}(\delta)$. δ is a $*$-derivation if $\delta(x^*) = \delta(x)^*$ for all x in $\mathcal{D}(\delta)$.*

Example 4.2. Let $\mathfrak{A} = C[0, 1]$ with the sup norm and let $\delta(f) = f'$ on all functions f with continuous derivative.

Example 4.3. Let $\mathfrak{A} = C_0(\mathbb{R})$, the algebra of all complex-valued continuous functions on $\mathbb{R}$ vanishing at ∞. Define $\delta(f) = f'$ on $C_0^{(1)}(\mathbb{R})$, the subalgebra of functions whose derivative is back in $C_0(\mathbb{R})$.

Remark. Let $\{U_t : t \in \mathbb{R}\}$ be the group of unilateral shifts on $C_0(\mathbb{R})$, i.e., $U_t(f)(x) = f(x + t)$, then we see that δ is a *generator* in the sense that it is the derivative of $\{U_t\}$, i.e.,

$$\delta(f)(x) = \lim_{t \to 0} \frac{U_t(f)(x) - f(x)}{t}.$$

Here is the more formal definition of *generator*.

Definition 4.7. *Let $\{\alpha_t : t \in \mathbb{R}\}$ be a one-parameter group of $*$-automorphisms on a C^*-algebra $\mathfrak{A}$, continuous in a suitable topology (often the strong or the norm topology), i.e., α_t is an automorphism for each t and we have the group property $\alpha_t \circ \alpha_s = \alpha_{t+s}$ for all s and t in $\mathbb{R}$. Then a derivation δ is said to be the generator of $\{\alpha_t : t \in \mathbb{R}\}$ if its domain $\mathcal{D}(\delta)$ coincides with the $*$-subalgebra of $\mathfrak{A}$ consisting of all x in $\mathfrak{A}$ such that $\lim\limits_{t\to 0} \frac{\alpha_t(x)-x}{t}$ exists, and $\delta(x)$ coincides with this limit for all x in $\mathcal{D}(\delta)$.*

As we have seen, the derivation in Example 4.3 is a generator. We shall see that the derivation in Example 4.1, on the other hand, is not! This will follow from the set of necessary and sufficient conditions discussed below in Theorem 4.9 for a derivation to be a generator of a one-parameter group of automorphisms of a C^*-algebra. For now let's show that the domain of definition of a generator cannot be too small.

Theorem 4.8. *Let $\{\alpha_t : t \in \mathbb{R}\}$ be a strongly continuous one-parameter group of automorphisms of a C^*-algebra $\mathfrak{A}$, with generator δ. Then the domain $\mathcal{D}(\delta)$ of δ is a strongly dense $*$-subalgebra of $\mathfrak{A}$.*

Proof. Let A be an element of $\mathfrak{A}$. For any $s > 0$ consider the integral $\frac{1}{s}\int_0^s \alpha_t(A)dt$. Then it is not too hard to show that the strong limit, as $s \to 0^+$ of these integrals is A itself. Moreover we can show that these integrals are in $\mathcal{D}(\delta)$, for (with s-lim denoting the strong limit)

$$\begin{aligned}
\operatorname*{s-lim}_{u\to 0} \frac{\left(\alpha_u\left(\frac{1}{s}\int_0^s \alpha_t(A)dt\right) - \frac{1}{s}\int_0^s \alpha_t(A)dt\right)}{u} &= \operatorname*{s-lim}_{u\to 0} \frac{1}{s}\frac{\left(\int_0^s \alpha_{t+u}(A) - \alpha_t(A)dt\right)}{u} \\
&= \operatorname*{s-lim}_{u\to 0} \frac{\left(\frac{1}{u}\int_u^{s+u} \alpha_t(A)dt - \frac{1}{u}\int_0^s \alpha_t(A)dt\right)}{s} \\
&= \frac{1}{s}(\alpha_s(A) - A).
\end{aligned}$$

□

A necessary condition for a derivation to be a generator is that it is closed in the appropriate topology. This means that if A_n is a sequence in $\mathcal{D}(\delta)$ converging to A and $\delta(A_n)$ converges to B, then $A \in \mathcal{D}(\delta)$ and $\delta(A) = B$.

The following theorem provides a set of necessary and sufficient conditions for a $*$-derivation on a C^*-algebra to be a generator of a strongly continuous one-parameter family of automorphisms.

Theorem 4.9 (BR, Theorem 3.2.51). *Let $\mathfrak{A}$ be a C^*-algebra with identity and let δ be a norm-densely defined norm-closed operator on $\mathfrak{A}$ with domain $\mathcal{D}(\delta)$. Then δ is the generator of a strongly continuous one-parameter group of $*$-automorphisms of $\mathfrak{A}$ if and only if the following hold:*

1. *$1 \in \mathcal{D}(\delta)$ and $\delta(1) = 0$.*
2. *$(1 \pm \delta)\mathcal{D}(\delta) = \mathfrak{A}$.*
3. *$\|(1 \pm \delta)A\| \geq \|A\|$ for all A in $\mathcal{D}(\delta)$.*

Proof. We will show that the three conditions are necessary for δ to be a generator and skip the more difficult direction which shows that they are sufficient (see [BR] for the full proof).

So we suppose that δ is the generator of a one-parameter group of a strongly continuous group of automorphisms $\{\alpha_t : t \geq 0\}$ on $\mathfrak{A}$. To see that $1 \in \mathcal{D}(\delta)$ note that $\alpha_t(1) = 1$ for all t and therefore $\lim_{t\to 0} \frac{\alpha_t(1)-1}{t} = 0$. This shows that $1 \in \mathcal{D}(\delta)$ and that $\delta(1) = 0$.

Next consider the so-called resolvent expression (note its resemblance to the Laplace transform)

$$R(A) = \int_0^\infty e^{-t}\alpha_t(A)dt$$

for any A in $\mathfrak{A}$. We shall show that $R(A) \in \mathcal{D}(\delta)$ and that $\delta(R(A)) = R(A) - A$. We have

$$\begin{aligned}
\lim_{s\to 0}\frac{\alpha_s(R(A)) - R(A)}{s} &= \lim_{s\to 0}\frac{\left(\int_0^\infty e^{-t}\alpha_{t+s}(A) - \int_0^\infty e^{-t}\alpha_t(A)dt\right)}{s} \\
&= \lim_{s\to 0}\frac{\left(\int_s^\infty (e^{-t+s}\alpha_t(A) - e^{-t}\alpha_t(A)dt - \int_0^s e^{-t}\alpha_t(A)dt\right)}{s} \\
&= \lim_{s\to 0}\left\{\left(\frac{e^s-1}{s}\right)\int_s^\infty e^{-t}\alpha_t(A)dt - \frac{\int_0^s e^{-t}\alpha_t(A)dt}{s}\right\} \\
&= R(A) - A.
\end{aligned}$$

Hence $(1-\delta)(R(A)) = A$. This shows that $(1-\delta)\mathcal{D}(\delta) = \mathfrak{A}$.

A similar proof shows that $(1+\delta)\mathcal{D}(\delta) = \mathfrak{A}$ as well.

To prove the last condition note that $R(A) = (1-\delta)^{-1}A$ and $\|R(A)\| \leq \int_0^\infty e^{-t}\|\alpha_t(A)\|dt = \|A\|\int_0^\infty e^{-t}dt = \|A\|$, so that $\|A\| \geq \|(1-\delta)^{-1}A\|$. □

To illustrate this theorem let's consider the three examples that have been presented so far.

Example 4.4. Consider the bounded derivation δ on $\mathfrak{A} = M_n(\mathbb{C})$ given by $\delta(y) = [ih, y]$, $y \in \mathfrak{A}$, where $h = h^*$.

1. Clearly $\delta(1) = 0$, in fact δ is 0 on any matrix that commutes with h.
2. $(1 \pm \delta)\mathfrak{A} = \mathfrak{A}$. Since $\mathfrak{A}$ is finite-dimensional it suffices to show that the linear maps $1 \pm \delta$ are one-one on $\mathfrak{A}$. We will show $1 - \delta$ is one-one. Suppose $a \in \mathfrak{A}$ and $(1-\delta)(a) = 0$, then

$$a = iha - iah.$$

Since h is hermitian we can find matrix units $\{e_{ij}\}$ of $\mathfrak{A}$ such that $ih = \sum_{k=1}^n \lambda_k e_{kk}$, for some $\lambda_k \in \mathbb{R}i$. Now suppose $a = \sum_{p,q=1}^n a_{pq}e_{pq}$, then

$$\begin{aligned}
\sum a_{pq}e_{pq} &= \sum \lambda_k e_{kk}\sum a_{pq}e_{pq} - \sum a_{pq}e_{pq}\sum \lambda_k e_{kk} \\
&= \sum a_{pq}(\lambda_p - \lambda_q)e_{pq}.
\end{aligned}$$

But this is a contradiction since the equation above says that $\lambda_p - \lambda_q$ is 1 whenever $a_{pq} \neq 0$. But $\lambda_p - \lambda_q \in \mathbb{R}i$, a contradiction.

3. We wish to show that $\|a\| \leq \|(1 \pm \delta)a\|$ for all $a \in \mathfrak{A}$. To do so it is enough to verify

$$\|a\| \leq \|(1 \pm \delta)a\|$$

for all $a \neq 0$ in $\mathfrak{A}$. To do this in the $(+)$ case it suffices to show

$$\|a^*a\| \leq \|(1+\delta)a^*(1+\delta)a\|.$$

Note that $(1+\delta)a^*(1+\delta)a = a^*a + \delta(a^*a) + \delta(a^*)\delta(a)$. Since a^*a is positive there exists a state ϕ such that $\phi(a^*a) = \|a^*a\|$. In fact note that ϕ can be taken to be the vector state $\langle \cdot\xi, \xi \rangle$ where ξ is an eigenvector corresponding to the largest eigenvalue λ of a^*a. Then $\phi(\delta(a^*a)) = \langle iha^*a\xi, \xi\rangle - \langle ia^*ah\xi, \xi\rangle = \lambda i(\langle h\xi, \xi\rangle - \langle h\xi, \xi\rangle) = 0$. Therefore

$$\|a\|^2 = \phi(a^*a) \leq \phi(a^*a + \delta(a^*a) + \delta(a^*)\delta(a)) \leq \|(1+\delta)a\|^2,$$

and we're done.

Example 4.5. $\delta f = f'$ on $C[0,1]$. Note that if $f(x) = e^x$ then $f \in \mathcal{D}(\delta)$ but $\delta f = f$ and therefore $\|(1-\delta)f\| \not\geq \|f\|$. Hence δ is not a generator.

Example 4.6. $\delta f = f'$ on its domain $C_0^{(1)}(\mathbb{R})$ in $\mathfrak{A} = C_0(\mathbb{R})$.

1. $\delta(1) = 0$ vacuously, or else we may adjoin the identity to the algebra $C_0(\mathbb{R})$ and define $\delta(1) = 0$.
2. We wish to show $(1 \pm \delta)\mathcal{D}(\delta) = \mathfrak{A}$. Let's treat the $(-)$ case. So if g is in $\mathfrak{A}$ we would like to try to solve

$$f' - f = g.$$

Using first order differential equations techniques we arrive at

$$f(t) = e^t\left(\int_0^t e^{-x}g(x)dx + c\right).$$

We need to show that there is a choice of c such that our solution is in $C_0(\mathbb{R})$. Pick c such that $\int_0^\infty e^{-x}g(x)dx + c = 0$. Then

$$\begin{aligned}\lim_{t\to\infty} e^t\left(\int_0^t e^{-x}g(x)dx + c\right) &= \lim_{t\to\infty} \frac{\int_0^t e^{-x}g(x)dx + c}{e^{-t}} \\ &= \lim_{t\to\infty} \frac{e^{-t}g(x)}{-e^{-t}} = 0.\end{aligned}$$

Now we look at the behavior of our solution for negative t. We'd like to show that our solution tends to 0 as $t \to -\infty$. Since $|\int_0^c e^{-x}g(x)dx| \leq \int_0^c e^{-x}|g(x)|dx$

we may assume without loss of generality that g is a non-negative function. If $\int_{-\infty}^{0} e^{-x} g(x)dx < \infty$ we're done. Otherwise note that

$$\lim_{t\to-\infty} e^t \int_0^t e^{-x} g(x)dx = \lim_{t\to-\infty} \frac{\int_0^t e^{-x} g(x)dx}{e^{-t}}$$

$$= \lim_{t\to-\infty} \frac{e^{-t} g(t)}{-e^{-t}} = 0.$$

3. We need to show that $\|(1 \pm \delta) f\| \geq \|f\|$ for all $f \in C_0^{(1)}(\mathbb{R})$. To do so pick a point k in $\mathbb{R}$ where $\|f\| = |f(k)|$. Then $|f(k)|^2$ is a maximum for $\bar{f} f$ so

$$\bar{f}'(k) f(k) + \bar{f}(k) f'(k) = 0.$$

Then

$$|f'(k) + f(k)|^2 = (\bar{f}'(k) + \bar{f}(k))(f'(k) + f(k))$$

$$= |f'(k)|^2 + \bar{f}'(k) f(k) + \bar{f}(k) f'(k) + |f(k)|^2$$

$$\geq \bar{f}(k) f(k) = \|f\|^2.$$

5. V: E_0-semigroups on $\mathcal{B}(\mathcal{H})$

In this lecture we will try to draw on our previous lectures to introduce the topic of E_0-semigroups. Our goal is to introduce the basic examples and to give some hints about the techniques that one uses to study them. The interested reader is encouraged to look up Powers' original paper on the subject, [Po4], where many of the important ideas in the subject are introduced. W. Arveson's book [A5] on E-semigroups is an invaluable resource for those who are interested in further study.

As mentioned in the first lecture E_0-semigroups arose from a question posed by Powers on the nature of dynamical systems on $\mathcal{B}(\mathcal{H})$ which proceed forward but not backward in time. It has turned out that this is a very difficult question but a very interesting one. As one might guess from the previous lectures the study relies rather heavily on unbounded symmetric operators, on derivations, and on semigroups. At this point we have enough background information to study the CAR flows which were studied by Powers in the 1980s.

Definition 5.1. *An E_0-semigroup $\alpha = \{\alpha_t : t \geq 0\}$ on $\mathcal{B}(\mathcal{H})$ is a one-parameter semigroup of endomorphisms α_t such that*

1. $\alpha_t(I) = I$,
2. $\alpha_t \circ \alpha_s = \alpha_{t+s}$ *for all* $s, t \geq 0$,
3. *the mapping* $t \to \alpha_t$ *is continuous in the weak-operator topology, i.e., for each* $A \in \mathcal{B}(\mathcal{H})$ *and for each* $f, g \in \mathcal{H}$, $t \to (\alpha_t(A) f, g)$ *is a continuous function of* t.

For us the interesting cases are those for which α_t is not onto for some $t > 0$ (and hence for all $t > 0$). For if α_t is onto $\mathcal{B}(\mathcal{H})$ for some t then α_t is onto for all $t > 0$ and then we can extend α for all t to a one-parameter group of $*$-automorphisms of $\mathcal{B}(\mathcal{H})$. This puts us in the framework of Wigner's result from the 1930s which shows that there is a one-parameter group of unitary operators $\{U_t : t \in \mathbb{R}\}$ such that the semigroup is of the form $\alpha_t(A) = U_t A U_t^*$ for all A in $\mathcal{B}(\mathcal{H})$.

There is a characterization of E_0-semigroups in terms of generators, just as there is for $\sigma-$weakly continuous groups of $*$-automorphisms on von Neumann algebras. (Compare the result below to Theorem 4.9.) All that we need to do is to sacrifice one of the range conditions.

Theorem 5.2. *Let $\mathcal{M}$ be a von Neumann algebra and suppose δ is a $*$-derivation on $\mathcal{M}$ whose domain $\mathcal{D}(\delta)$ is weak* dense in $\mathcal{M}$. Then δ is the generator of a σ-weakly continuous one-parameter semigroup of $*$-automorphisms on $\mathcal{M}$ if and only if the following hold.*

1. $\delta(I) = 0$.
2. $(I \pm \delta)\mathcal{D}(\delta) = \mathcal{M}$.
3. $\|(I \pm \delta)A\| \geq \|A\|$, *all* $A \in \mathcal{D}(\delta)$.

If one replaces 2) *with* 2'), $(I + \delta)\mathcal{D}(\delta) = \mathcal{M}$, *then δ is the generator of an E_0-semigroup.*

Here is the main example which we propose to discuss in this lecture, the CAR flow.

Definition 5.3. *The CAR Flow Let $\mathfrak{A}$ be the CAR algebra over a Hilbert space $\mathcal{H}$ with corresponding antisymmetric Fock space $\mathcal{F}(\mathcal{H})$. Recall that $\mathfrak{A}$ is generated by the operators $a(f)$, $f \in \mathcal{H}$. Define $\alpha_t(a(f)) = a(S_t f)$, where S_t is the unilateral shift.*

Remark. Let's make some calculations that suggest at least that our definition actually gives us an E_0-semigroup. To see 1) note that $I = a(f)^*a(f) + a(f)a(f)^* = \{a(f)^*, a(f)\}_+$ for any unit vector, so $\alpha_t(I) = \{a(S_t)^*, a(S_t f)\}_+$, so $\alpha_t(I) = I$. To see that 2) holds note that $\alpha_t(\alpha_s(a(f))) = \alpha_t(a(S_t f)) = a(S_t(S_s f)) = a(S_{t+s} f)$, giving us the semigroup property. Finally 3) follows from the fact that $t \to S_t$ is strongly continuous on $\mathcal{H}$.

Before we continue with the CAR flow one might ask why we are content to limit our study of E_0-semigroups to the type I factors. The answer is that the subject is already extremely difficult in this setting! Although interesting work has been done by A. Alevras and R. Powers on the hyperfinite type II_1 case, calculations become more difficult there from the fact that pairs of units (see below) do not act as nicely together as they do in the type I case.

Recall that an *isometry* is a bounded linear map V on a Hilbert space $\mathcal{H}$ with the property that $\|Vf\| = \|f\|$ for all $f \in \mathcal{H}$.

Definition 5.4. *Let α be an E_0-semigroup on $\mathcal{B}(\mathcal{H})$. A unit is a strongly continuous one-parameter family of isometries $\{U_t : t \geq 0\}$ with the property that*

1. $U_{t+s} = U_t U_s$, *for all* $s, t \in \mathbb{R}^+$.
2. $U_t^* U_t = I$ *(this is the isometric property).*
3. $U_t A = \alpha_t(A) U_t$ *for all* $A \in \mathcal{B}(\mathcal{H})$ *and* $t \geq 0$.

The most familiar one-parameter family of isometries is the family of unilateral shifts $S_t,\ t \geq 0$ on $L^2(0,\infty)$, which we have looked at before. Let's show that there is a unit of the CAR flow which is closely related to $\{S_t : t \geq 0\}$. Consider $\mathcal{F} = \mathcal{F}(\mathcal{H})$, the antisymmetric Fock space over $\mathcal{H} = L^2(0,\infty)$. We shall define a semigroup of isometries $U_t,\ t \geq 0$ acting on $\mathcal{F}$ as follows:

$$U_t\Omega = \Omega, \text{ and}$$

$$U_t(f_1 \wedge f_2 \wedge \cdots \wedge f_n) = S_t f_1 \wedge S_t f_2 \wedge \cdots \wedge S_t f_n, \text{ where}$$

$f_1, f_2, \ldots, f_n$ is an arbitrary list of vectors in $\mathcal{H}$. Now consider the following computation: $U_t(a(f)(f_1 \wedge f_2 \wedge \cdots \wedge f_n)) = U_t(f \wedge f_1 \wedge \cdots \wedge f_n) = S_t f \wedge S_t f_1 \wedge \cdots \wedge S_t f_n$, and $\alpha_t(a(f)*)(U_t(f_1 \wedge \cdots \wedge f_n)) = a(S_t f)^*(S_t f_1 \wedge \cdots \wedge S_t f_n) = S_t f \wedge S_t f_1 \wedge \cdots \wedge S_t f_n$, and so we have some compelling evidence that $U_t, t \geq 0$ is a unit for α, i.e., $U_t A = \alpha_t(A)U_t$.

Are there any other units for the CAR flow? The answer is yes, and a nice characterization of them was provided by B. V. R. Bhat, in [Bh1]. Although we will not have enough time to go into details, we do wish to discuss the notion of the index of an E_0-semigroup, which is related to the "number" of units that an E_0-semigroup has.

In order to begin to study the index, let's consider a pair of units $U_t,\ t \geq 0$ and $V_t,\ t \geq 0$, for any E_0-semigroup on $\mathcal{B}(\mathcal{H})$ ($\mathcal{H}$ is now any infinite-dimensional Hilbert space). From the definition of unit we have $U_t A = \alpha_t(A)U_t$ for all A and $t \geq 0$. Taking adjoints we have $A^*U_t^* = U_t^*\alpha_t(A)^* = U_t^*\alpha_t(A^*)$. Replacing A with its adjoint gives us $AU_t^* = U_t^*\alpha_t(A)$. Then since U_t and $V_t,\ t \geq 0$, are both units we have

$$U_t^* V_t A = U_t^* \alpha_t(A) V_t = A U_t^* V_t,$$

so that the operators $U_t^* V_t$ commute with everything in $\mathcal{B}(\mathcal{H})$! Since the scalar multiples of the identity operator are the only members of the commutant of $\mathcal{B}(\mathcal{H})$ we conclude that $U_t^* V_t$ is a scalar. Moreover, for non-negative s and t,

$$\begin{aligned}(U_t^* V_t)(U_s^* V_s) &= U_t^* U_s^* \alpha_s(V_t) V_s \\ &= U_t^* U_s^* V_s V_t \\ &= (U_s U_t)^* V_s V_t \\ &= U_{s+t}^* V_{s+t}\end{aligned}$$

so that $\{U_t^* V_t\}$ is a continuous semigroup of scalar operators and therefore there is a complex number λ which satisfies

$$U_t^* V_t = e^{-\lambda t} I.$$

Modifying one of the units through multiplying it by a semigroup of scalar operators $e^{it\theta} I$, if necessary, we may assume that our scalar λ is a positive real number. So we have shown the following.

Lemma 5.5. *Suppose U_t and V_t are a pair of distinct nontrivial (i.e., not scalar) units for an E_0–semigroup α then there is a $\lambda \in \mathbb{C}$ with positive real part such that $U_t^* V_t =$*

$e^{-\lambda t}I$. *By modifying one of the units through multiplication by a trivial unit we may assume that* λ *is in fact real.*

Now we plan to study the generators of U_t and V_t to see what more we can discover. Let D, (respectively, d) be the generator of $\{V_t\}$, (respectively, of $\{U_t\}$). Since both semigroups are semigroups of isometries it turns out that both D and d are maximal symmetric linear operators with deficiency indices $(0, \infty)$ and with dense domains $\mathcal{D}(D)$, $\mathcal{D}(d)$ in $\mathcal{H}$.

What does this mean in terms of the symmetric operators we have already studied? It means that we would like to assume that one of the generators, say d, is the direct sum of infinitely many copies of the operator $iA_4 = -d/dx$ which we encountered in Example 4 in the lectures on symmetric operators and deficiency indices. If we go ahead and make this assumption then we should note that since A_4 is symmetric it follows that d is skew-symmetric, i.e., $(df, g) + (f, dg) = 0$ for all f, g in $\mathcal{D}(d)$ because $(iA_4 f, g) + (f, iA_4 g) = 0$ for all f, g in $\mathcal{D}(A_4)$.

Expressing things in another way, we would like to assume for convenience and for the sake of familiarity that our Hilbert space $\mathcal{H}$ is (or is unitarily equivalent to) the direct sum of infinitely many copies of $L^2(0, \infty)$, where $\mathcal{D}(d)$ consists of infinitely many copies of functions $f(x) = f_1(x) \oplus f_2(x) \oplus \cdots$ where both $\sum_{j=1}^{\infty} \|f_j\|^2$ and $\sum_{j=1}^{\infty} \|f_j'\|^2$ converge, and

$$df(x) = -(f_1'(x) \oplus f_2'(x) \oplus \cdots).$$

This is not such a terrible thing to do because any symmetric operator of deficiency indices $(0, \infty)$ is actually unitarily equivalent to infinitely many copies of A_4, [RSz]. Let's express this in a final and less cumbersome way. We will view the Hilbert space $\mathcal{H}$ as $\mathcal{H} = \mathcal{K} \otimes L^2(0, \infty)$ where $\mathcal{K}$ is an infinite-dimensional Hilbert space and $\mathcal{D}(d)$ consists of all differentiable functions in $\mathcal{H}$ which are 0 at the origin, i.e., $f(0) = 0 \in \mathcal{K}$. Then $d \subset -d^*$ and d^* is the operator d/dx on its domain consisting of all differentiable functions f in $\mathcal{H}$ with arbitrary boundary conditions $f(0) \in \mathcal{K}$. So what is $\mathcal{D}(D)$?

Let $f \in \mathcal{D}(D) \subset \mathcal{H} = \mathcal{K} \otimes L^2(0, \infty)$ and $g \in \mathcal{D}(d)$. Then

$$\begin{aligned}(V_t f, U_t g) &= (U_t^* V_t f, g)\\ &= e^{-\lambda t}(f, g),\end{aligned}$$

so taking d/dt at $t = 0$ on both sides of this equation yields

$$\begin{aligned}-\lambda(f, g) &= (Df, g) + (f, dg), \text{ or}\\ ((-D - \lambda I)f, g) &= (f, dg) = -(d^* f, g), \text{ which implies}\\ (D + \lambda I)f &= d^* f.\end{aligned}$$

Caution: this last equation does *not* mean that $D + \lambda I$ is equal to d^* but it *does* mean that $D + \lambda \subset d^*$, or in other words, that $D + \lambda$ agrees with the operator d^* on the restricted domain $\mathcal{D}(D)$. So we conclude that $\mathcal{D}(D)$ consists of some linear submanifold of the set of all differentiable functions in $\mathcal{K} \otimes L^2(0, \infty)$ with arbitrary boundary conditions (see the discussion about A_4^* in the previous lectures).

The following lemma gives more insight into the domain $\mathcal{D}(D)$ of D.

Lemma 5.6. *Suppose V_t and U_t are distinct units of α satisfying $U_t^* V_t = e^{-\lambda t} I$ with $\lambda > 0$. Then $\mathcal{D}(D) \cap \mathcal{D}(d) = \{0\}$.*

Proof. Let $f \in \mathcal{K} \otimes L^2(0, \infty)$ be in both $\mathcal{D}(D)$ and $\mathcal{D}(d)$ then by taking the derivative of the identity $\|f\|^2 = (V_t f, V_t f)$ at $t = 0$ we get, since $d \subset -d^*$,

$$\begin{aligned} 0 &= (Df, f) + (f, Df) \\ &= ((d^* - \lambda) f, f) + (f, (d^* - \lambda) f) \\ &= (-df, f) + (f, -df) - (\lambda f, f) - (f, \lambda f) \\ &= 0 - 2(\lambda f, f) \\ &= -2\lambda \|f\|^2, \end{aligned}$$

a contradiction unless $f = 0$. □

Suppose f and g are in $\mathcal{D}(d^*)$: then by integration by parts,

$$\begin{aligned} (d^* f, g) + (f, d^* g) &= (f', g) + (f, g') \\ &= -\langle f(0), g(0) \rangle \end{aligned}$$

where the last inner product takes place in $\mathcal{K}$, which we are identifying with the boundary of $\mathcal{H} = \mathcal{K} \otimes L^2(0, \infty)$. Using this identity we come to the following characterization of the domain $\mathcal{D}(D)$ of D.

Lemma 5.7. *Let U_t and V_t be units of α satisfying $U_t^* V_t = e^{-\lambda t} I$ for all $t \geq 0$. Let d and D be their generators, respectively, on the Hilbert space $\mathcal{H} = \mathcal{K} \otimes L^2(0, \infty)$. Assume that d is the skew-symmetric operator $-d/dx$ whose domain $\mathcal{D}(d)$ consists of all differentiable functions f in $\mathcal{H}$ such that $f(0) = 0$. Then there exists an isometry V from $\mathcal{H}$ to its boundary $\mathcal{K}$ such that $\mathcal{D}(D) = \{f \in \mathcal{D}(d^*) : (Vf)(0) = \sqrt{\lambda} f(0)\}$.*

Proof sketch. D is skew-symmetric, i.e., for f and g in $\mathcal{D}(D)$, $(Df, g) + (f, Dg) = 0$. This follows by taking the derivative of the expression $(f, g) = (V_t f, V_t g)$ at $t = 0$ as we did in the previous lemma. Then

$$\begin{aligned} 0 = (Df, g) + (f, Dg) &= ((-d^* - \lambda I) f, g) + (f, (-d^* - \lambda I) g) \\ &= \langle f(0), g(0) \rangle - 2\lambda (f, g) \\ &= \langle f(0), g(0) \rangle + (\sqrt{2\lambda} f, \sqrt{2\lambda} g). \end{aligned}$$

Hence there is an isometry V from $\mathcal{D}(D)$ to $\mathcal{K}$ satisfying $Vf = \frac{1}{\sqrt{2\lambda}} f(0)$. By continuity the isometry extends to all of $\mathcal{H}$. □

A more complicated (but similar) argument to the two previous lemmas proves the following lemma. Details may be found in [PP].

Lemma 5.8. *Let α be an E_0-semigroup and suppose U_t, $V_t^{(1)}$ and $V_t^{(2)}$ are units of α. Assume that $\mathcal{H}$ is identified with $\mathcal{K} \otimes L^2(0, \infty)$ as above. Let $V^{(1)}$ and $V^{(2)}$ be the isometries characterizing the domains of the generators D_1 and D_2 of the semigroups of isometries $V_t^{(1)}$ and $V_t^{(2)}$ as in the previous lemma. Then $V^{(1)*} V^{(2)}$ is a scalar multiple of the identity.*

The previous results lead to the following important theorem in the subject, which is due to W. Arveson, although in the form it is expressed below it is to be found in [PP].

Theorem 5.9. *Let α be an E_0-semigroup and consider the family of units which have real covariance with U_t, i.e., $U_t^* V_t = e^{-\lambda t}$ for all $t \geq 0$. Consider the the corresponding isometries from $\mathcal{H}$ to $\mathcal{K}$ associated with the units. Then these isometries form a Hilbert space under the inner product $(V^{(1)}, V^{(2)}) = V^{(2)*} V^{(1)} I$.*

Now we are able to introduce the notion of the index of an E_0-semigroup. This notion has undergone several evolutions. The one presented below is a bit of an amalgamation of the original approaches of Arveson and of Powers.

Definition 5.10. *The index of an E_0-semigroup is the dimension of the Hilbert space generated by the isometries in the previous theorem.*

The index is an important cocycle conjugacy invariant. We define cocycle conjugacy below, which is related to an equivalence relation on automorphisms considered by A. Connes in his classification of the automorphisms on the hyperfinite II_1 factor.

Definition 5.11. *A pair of E_0-semigroups α and β on $\mathcal{B}(\mathcal{H})$ are said to be conjugate if there is an automorphism γ of $\mathcal{B}(\mathcal{H})$ such that $\alpha_t \circ \gamma = \gamma \circ \beta_t$ for all $t \geq 0$. An E_0-semigroup α' is a cocycle perturbation of α if there is a strongly continuous group $U = \{U_t : t \geq 0\}$ of unitary cocycles (i.e., $U_{t+s} = U_t \alpha_t(U_s)$ for all non-negative s and t) such that $\alpha'_t(A) = U_t \alpha_t(A) U_t^*$ for all A in $\mathcal{B}(\mathcal{H})$. α and β are cocycle conjugate if β is conjugate to a cocycle perturbation α' of α.*

Theorem 5.12. *The index $ind(\alpha)$ of an E_0-semigroup is a cocycle conjugacy invariant.*

Using his approach to the index Arveson was able to establish the following important theorem.

Theorem 5.13 (A1). *Let α and β be E_0-semigroups on $\mathcal{B}(\mathcal{H}_\alpha)$ and $\mathcal{B}(\mathcal{H}_\beta)$ respectively. Then the index of the E_0-semigroup $\gamma = \alpha \otimes \beta$ on $\mathcal{B}(\mathcal{H}_\alpha \otimes H_\beta)$ is the sum $ind(\alpha) + ind(\beta)$.*

If we return to the CAR flow we can get a feeling perhaps for what is happening in the result above. Recall that we constructed the CAR algebra over the Hilbert space $L^2(0, \infty)$. It takes some doing but one can show (and Powers was the first to show) that this E_0-semigroup has index 1. Now suppose we construct the Hilbert space $\mathfrak{H}$ consisting of n copies of $L^2(0, \infty)$ (n could be ∞!). Then one can construct the CAR relations over this Hilbert space just as one did for a single copy of $L^2(0, \infty)$ and one can show that the CAR algebra over $\mathfrak{H}$ gives us a representation on the corresponding antisymmetric Fock space $\mathcal{F}(\mathfrak{H})$. The (generalized) CAR flow in this case is obtained by sending $a(f)$, $f = \oplus_{j=1}^n f_j$ where each f_j is a function in $L^2(0, \infty)$, to $\mathcal{S}_t f = \oplus_{j=1}^n S_t f_n$. Then it is possible to show (although we shall not do so in these notes) that $\mathcal{B}(\mathcal{F}(\mathfrak{H}))$ is isomorphic to $\otimes_{j=1}^n \mathcal{B}(\mathcal{F}(\mathcal{H}))$ and that the CAR flow is isomorphic to the tensor product of n copies of the original CAR flow on $\mathcal{B}(\mathcal{F}(\mathcal{H}))$. Therefore it is at least plausible (and it's true) that the index of the CAR flow in question is n. Powers was the first to show this and Arveson later generalized this result, [Po4, A1].

Why did Powers' result require a generalization? The answer is because not every E_0−semigroup is constructed like the ones above. In fact, Powers showed that there are examples of E_0-semigroups that have no units whatsoever ([Po6])! This leads us to consider next the notion of the "type" of an E_0-semigroup, which is determined by properties of its units. In order to discuss the type of an E_0-semigroup it is useful first to examine Arveson's product systems, which we shall leave to the next lecture.

6. More recent developments

In this section we discuss some of the developments in the theory about those E_0-semigroups that are not cocycle conjugate to the CAR flows. We begin with a description of product systems associated with E_0-semigroups. As Arveson and others have shown, product systems provide a powerful tool for proving results about E_0-semigroups. Arveson's proof of the additivity of the index under tensor product (cf. Theorem 5.13) depends on techniques using product systems, for example. As we shall discuss below, Arveson has shown that every concrete product system may be realized as the product system of an E_0-semigroup. This result may be combined with B. Tsirelson's results on product systems to deduce the existence of uncountably many equivalence classes of E_0-semigroups which are not cocycle conjugate to the CAR flows defined above.

Given an E_0-semigroup α its corresponding product system $\mathcal{E}_\alpha$ may be described as follows [A5]. For fixed $t > 0$ let $\mathcal{E}_\alpha(t)$ be the linear space consisting of all T in $\mathcal{B}(\mathcal{H})$ satisfying the identities

$$TA = \alpha_t(A)T,$$

for all $A \in \mathcal{B}(\mathcal{H})$. By taking adjoints and using the fact that α is a $*$-endomorphism the identities above imply also that $S^*\alpha_t(A) = AS^*$ for all $A \in \mathcal{B}(\mathcal{H})$. Combining these identities shows that $S^*T \in \mathcal{B}(\mathcal{H})' \cap \mathcal{B}(\mathcal{H})$, i.e., that S^*T commutes with all of $\mathcal{B}(\mathcal{H})$, and therefore S^*T is a scalar multiple of the identity. This observation shows that for each $t > 0$ $\mathcal{E}_\alpha(t)$ becomes a Hilbert space with inner product $\langle T, S\rangle = S^*T$. If $T \in \mathcal{E}_\alpha(t)$ and $S \in \mathcal{E}_\alpha(s)$ it is easy to see that $ST \in \mathcal{E}_\alpha(s+t)$. In this sense the family $\mathcal{E}_\alpha = \{\mathcal{E}_\alpha(t) : t \geq 0\}$ acts like a continuous tensor product. In fact, it can be shown that $\mathcal{E}_\alpha(s+t)$ is the norm closed linear span of $\mathcal{E}_\alpha(s)\mathcal{E}_\alpha(t)$.

What is the relationship between the product systems of a pair of cocycle conjugate E_0-semigroups? They are the same, up to isomorphism. In fact, Arveson has shown [A1] that α and β are cocycle conjugate if and only if $\mathcal{E}_\alpha$ and $\mathcal{E}_\beta$ are isomorphic in a natural sense. Even more striking is the following very important result of Arveson [A2] (see also [A6] and [Sk] for shorter versions of the proof). First we require the definition of a concrete product system [A5, sect. 2.4].

Definition 6.1. *Let $\mathcal{H}$ be a separable Hilbert space and let $\mathcal{E}$ be a Borel subset of $(0,\infty)\times\mathcal{B}(\mathcal{H})$. Then $\mathcal{E}$ is said to be a concrete product system if it satisfies the following properties.*

1. *For all $t > 0$, $\mathcal{E}(t) = \{T : T \in \{t\} \times \mathcal{B}(\mathcal{H})\} \cap \mathcal{E}$ is a norm closed linear subspace of $\mathcal{B}(\mathcal{H})$ such that S^*T is a scalar for all S and T in $\mathcal{E}(t)$.*

2. *$\mathcal{E}(s+t)$ is the norm closed linear span of $\mathcal{E}(s)\mathcal{E}(t)$, the set of all products ST with S in $\mathcal{E}(s)$ and T in $\mathcal{E}(t)$.*
3. *$\mathcal{E}(s)$ is isomorphic as a measurable family of Hilbert spaces to $(0, \infty) \times H_0$ where H_0 is a separable Hilbert space.*

In the statement of Arveson's theorem below the term *abstract product system* refers to a family of separable family of Hilbert spaces $E(t)$ for $t \geq 0$ with bilinear maps $(x, y) \in E(s) \times E(t) \to xy \in E(s+t)$ which acts like tensoring. We will not present the precise definition of an abstract product system. There are some very important consequences of the theorem, which we shall touch upon below.

Theorem 6.2 (A5). *Every abstract product system $\mathcal{E}$ is isomorphic (in a natural sense) to the concrete product system of an E_0-semigroup.*

Now we come to the definition of the *type* of an E_0-semigroup, which is a cocycle conjugacy invariant. Let $\mathcal{U}_\alpha$ be the set of all units of an E_0-semigroup α on $\mathcal{B}(\mathcal{H})$. For fixed $t > 0$ consider all products of the form $T = U_1(t_1)U_2(t_2)\cdots U_n(t_n)$ where $t_1, t_2, \ldots, t_n$ are non-negative real numbers such that $t_1 + \cdots + t_n = t$ and $U_1, U_2, \ldots, U_n$ are (not necessarily distinct) units in $\mathcal{U}_\alpha$. Note that $T \in \mathcal{E}_\alpha(t)$. Then we say that α is of Type *I* if $\mathcal{E}_\alpha(t)$ is the linear span of all such operators T. (In this case we also say that α is completely spatial.) It turns out that the CAR flows discussed above are completely spatial. In fact, Arveson has obtained the following characterization of completely spatial [A2], (see also [Za]).

Theorem 6.3. *An E_0-semigroup is completely spatial if and only if it is cocycle conjugate to a CAR flow.*

Using this result it turns out that the index of a type *I* E_0-semigroup is a complete conjugacy invariant. Are there E_0-semigroups which are not of type *I*? Through a very complicated construction Powers in [Po6] produced an example of an E_0-semigroup which has no units at all (see also Arveson's description of Powers' result in [A5, Chap. 13]). Powers' construction is carried out on the CAR algebra. An important component of the proof involves the work of Powers and Størmer on quasifree states on the CAR algebra, [PS]. Using another very technical construction Powers produced [Po7] an example of a spatial E_0-semigroup (i.e., it has units) which is not completely spatial. E_0-semigroups that are spatial but not completely spatial are said to be of type *II*. E_0-semigroups without units are called type *III*. For spatial E_0-semigroups, α is of type I_n or II_n. This means, in a sense which Arveson has made precise, that there are $n+1$ units for α.

B. Tsirelson has made extremely significant contributions to the theory of E_0-semigroups of type *II* and *III*. Using probabilistic methods he shows [T] that there are uncountably many non-isomorphic product systems which correspond (after an application of Arveson's Theorem 6.2 above) to E_0-semigroups of type *III*. Thus there are uncountably many E_0-semigroups of type *III* (up to cocycle conjugacy). In the same article Tsirelson produces product systems which give rise to E_0-semigroups of type II_0 (i.e., they have only one unit up to perturbation by trivial scalar-valued semigroups). This result provides an affirmative answer to the question of the existence of type II_0 E_0-semigroups. In fact, Tsirelson's work shows that there are uncountably many of these, up to cocycle conjugacy.

In recent years completely positive semigroups (or CP-semigroups) have played an increasingly important role in the theory. This is largely due to the fundamental result of R. Bhat [Bh3] that every contractive CP-semigroup "dilates" to an E-semigroup as well as to the fact that important invariants associated with a CP-semigroup are preserved under dilation. In particular, every unital CP-semigroup dilates to an E_0-semigroup. We shall discuss the notion of dilation below. First we shall recall Stinespring's result on the dilation of a single completely positive map.

Definition 6.4. *A linear map ϕ from a C^*-algebra $\mathfrak{A}$ to $\mathcal{B}(\mathcal{H})$ is completely positive if for all $n \in \mathbb{N}$, $x_1, x_2, \ldots, x_n$ in $\mathfrak{A}$ and $y_1, y_2, \ldots, y_n$ in $\mathcal{B}(\mathcal{H})$,*

$$\sum_{i=1}^{n}\sum_{j=1}^{n} y_i^*\phi(x_i^*x_j)y_j \geq 0.$$

Note from the definition that completely positive implies positive as $\phi(x^*x) \geq 0$ for all $x \in \mathfrak{A}$.

Stinespring showed [Sti] that every CP-map can be dilated to a $*$-representation. This means that there is a Hilbert space $\mathcal{K}$, an isometry V from $\mathcal{H}$ to $\mathcal{K}$, and a $*$-representation π from $\mathfrak{A}$ to $\mathcal{B}(\mathcal{K})$ such that for all $a \in \mathfrak{A}$,

$$\phi(a) = V^*\pi(a)V.$$

A CP-semigroup on $\mathcal{B}(\mathcal{H})$ is a one-parameter semigroup $\phi = \{\phi_t : t \geq 0\}$ of CP-maps from $\mathcal{B}(\mathcal{H})$ into itself such that the maps $t \to \langle f, \phi_t(a)g\rangle$ is continuous in t for all $a \in \mathcal{B}(\mathcal{H})$ and f, g in $\mathcal{H}$. The following is Bhat's theorem on dilations ([Bh3] (cf. also [A5, Chap. 8] for an elegant proof of this result.

Theorem 6.5. *Let ϕ be a unital CP-semigroup on $\mathcal{B}(\mathcal{H})$. Then there is a Hilbert space $\mathcal{K}$, an E_0-semigroup α on $\mathcal{B}(\mathcal{K})$, and an isometry W from $\mathcal{H}$ to $\mathcal{K}$ such that $\alpha_t(WW^*) \geq WW^*$ so that*

$$\phi_t(A) = W^*\alpha_t(WAW^*)W$$

for all $A \in \mathcal{B}(\mathcal{H})$ and all $t \geq 0$.

There is a sense in which α can be chosen to be minimal among all dilations. This simply means that the set

$$\{\alpha_{t_1}(WA_1W^*)\alpha_{t_2}(WA_2W^*)\cdots\alpha_{t_n}(WA_nW^*)f : t_i \geq 0, f \in \mathcal{H}, A_i \in \mathcal{B}(\mathcal{H})\}$$

generates $\mathcal{K}$. Minimal dilations are unique up to conjugacy.

It is an interesting problem to try to determine the type and index of a minimal dilation of a CP-semigroup. Consider, for example, the CP-semigroup on $M_2(\mathbb{C})$ defined by $\phi_t\begin{pmatrix} a & b \\ c & d\end{pmatrix} = \begin{pmatrix} a & e^{-t}b \\ e^{-t}c & d\end{pmatrix}$. Note that ϕ is unital. The minimal dilation of ϕ is of type I_1. In fact Arveson and Powers independently showed that the minimal dilations of all unital CP-semigroups on $M_n(\mathbb{C})$ are of type I, [A7], [Po5]. In [A7] Arveson has also developed the notion of units for CP-semigroups which leads in turn to a numerical index. He has shown that this index corresponds to the index of the minimal dilation (cf. also [Bh2] for results on the index of a CP-semigroup).

Definition 6.6. *Let ϕ be a CP-semigroup on $\mathcal{B}(\mathcal{H})$. A CP-semigroup ϕ' on $\mathcal{B}(\mathcal{H})$ is subordinate to ϕ if the linear mappings $\phi_t - \phi'_t$ are completely positive for all $t \geq 0$.*

The subordinates of an E_0-semigroup form a lattice. Arveson showed that an order isomorphism exists between the subordinates of a CP- semigroup on a matrix algebra and the CP-semigroups of its minimal dilation. In [Bh1] and [Po2] Bhat and Powers extended this result to all unital CP-semigroups and their minimal dilations.

Powers has developed a powerful technique to study spatial E_0-semigroups arising as the minimal dilations of CP-semigroups called CP-flows. In particular this construction yields uncountably many E_0-semigroups of type II_0. He has also developed a criterion whereby one can determine whether a pair of E_0-semigroups obtained this way are cocycle conjugate. We touch only briefly upon a special case of Powers' work to try to convey a feeling for the results he has obtained.

Let $\mathcal{H}$ be the Hilbert space $\mathcal{K} \otimes L^2(0, \infty)$ where $\mathcal{K}$ is a Hilbert space, i.e, $\mathcal{H}$ is the Hilbert space of all measurable functions f such that $f(t) \in \mathcal{K}$ for all $t \geq 0$ and $\int_0^\infty \|f(t)\|^2 dt < \infty$. Consider the unilateral shift semigroup of isometries $S_t, t \geq 0$ on $\mathcal{H}$ given by

$$(S_t f)(x) = \begin{cases} 0, & \text{for} \quad 0 \leq x < t, \\ f(x-t), & \text{for} \quad x \geq t. \end{cases}$$

Definition 6.7. *A CP-flow over $\mathcal{K}$ CP-semigroup on $\mathcal{H} = \mathcal{K} \otimes L^2(0, \infty)$ which satisfies $S_t A = \alpha_t(A) S_t$ for all $t \geq 0$ and all $A \in \mathcal{B}(\mathcal{H})$.*

We consider the case where $\dim(\mathcal{K}) = 1$, i.e., where $\mathcal{H} = L^2(0, \infty)$ and where some of the concepts appearing in Powers' work are easier to define. Let Λ be the multiplication operator on $\mathcal{H}$ defined by $(\Lambda f)(x) = e^{-x} f(x)$. A *boundary weight* ω on $\mathcal{B}(\mathcal{H})$ is a positive weight with the property that either ω is bounded or it is unbounded only at the origin. More specifically there is a positive linear function μ in $\mathcal{B}(\mathcal{H})_*$, the predual of $\mathcal{B}(\mathcal{H})$, where μ is defined by

$$\mu(A) = \omega((I - \Lambda)^{-\frac{1}{2}} A (I - \Lambda)^{-\frac{1}{2}})$$

for all $A \in \mathcal{B}(\mathcal{H})$. A boundary weight is said to be normalized if it satisfies $\omega(I - \Lambda) = 1$ (note that this makes μ a state). In [Po1] Powers has shown how to assign a unital CP-flow α to each normalized boundary weight. Let α^d be the minimal dilation of α to an E_0-semigroup.

Theorem 6.8. *Let ω be a normalized boundary weight on $\mathcal{B}(\mathcal{H})$.*

1. *Suppose ω is bounded, i.e., $\omega(I) < \infty$. Then α^d is an E_0-semigroup of type I_n, where n corresponds to the rank of ω, i.e., if $\omega(A) = \sum_{k=1}^n \langle f_k, A f_k \rangle$ where $f_1, f_2, \ldots$ are linearly independent vectors in $\mathcal{H}$, then the rank of ω is n (which could be ∞).*
2. *If ω is unbounded then α^d is an E_0-semigroup of type II_0.*

In [APP] A. Alevras, Powers and Price studied the structure of the family of type II_0 E_0-semigroups that arise from unbounded boundary weights, as above. They have completely characterized the multiplicative structure of the local unitary cocycles of these type II_0 examples (a strongly continuous family U_t, $t \geq 0$ of unitary operators is called a local cocycle if $U_{t+s} = U_t\alpha_t(U_s)$ for all non-negative s and t and U_t lies in the commutant of the image of α_t^d for each t). This result provides a window into the structure of type II_0 examples, about which very little had been known previously.

In these notes we have elected to use the CAR flows in order to describe the type I examples (Theorem 5.13) but there are other ways to approach the subject, including the use of CCR flows. In [PR] Powers and D. W. Robinson showed that the CAR and the CCR flows of index n are conjugate. In order to touch upon the recent work of M. Izumi [Iz] on E_0-semigroups of type I and type III we describe how a flow of type I can be constructed using a Weyl system of unitary operators. The reader is referred to [A5] for a detailed description of this construction. Given a Hilbert space $\mathcal{H}$ a Weyl system over $\mathcal{H}$ is a strongly continuous mapping from $\mathcal{H}$ to the unitary operators in $\mathcal{B}(\mathcal{K})$ for some Hilbert space $\mathcal{K}$, $\xi \to W_\xi$, where the unitary operators W_ξ satisfy the Weyl form of the commutation relations,

$$W_\xi W_\eta = e^{i\mathcal{J}\langle\xi,\eta\rangle} W_{\xi+\eta},$$

where $\mathcal{J}\langle\xi,\eta\rangle$ is the imaginary part of the inner product $\langle\xi,\eta\rangle$. If $\mathcal{K}$ is the symmetric Fock space over $\mathcal{H}$ (which we will not describe here) the linear span of the Weyl operators is strongly dense in $\mathcal{B}(\mathcal{K})$. Suppose $\mathcal{H} = \mathcal{K}_0 \otimes L^2(0,\infty)$, where $\mathcal{K}_0$ is a Hilbert space. Let the unilateral shifts S_t act on $\mathcal{H}$ as above. The mapping $W_\xi \to \alpha_t(W_\xi) = W_{S_t\xi}$ gives rise to an E_0-semigroup on $\mathcal{B}(\mathcal{K})$ called the CCR flow of index d where $d = \dim(\mathcal{K}_0)$.

Recently Izumi [Iz] has constructed a very interesting family of E_0-semigroups that are defined using the Weyl system with $\mathcal{H} = L^2(0,\infty)$ as above. Izumi has shown that this family includes all of Tsirelson's examples of E_0-semigroups of type III (cf. also [BhS] and [IzS]). Izumi's flows are obtained by constructing one-parameter semigroups $\{T_t : t \geq 0\}$ of bounded operators which satisfy the following two conditions:

C1 $\quad T_t^* S_t = I, \qquad t \geq 0,$

C2 $\quad T_t - S_t$ is a Hilbert Schmidt operator for all $t \geq 0$.

The associated E_0-semigroups are defined as follows. Let $L^2(0,\infty)_{\mathbb{R}}$ be the linear submanifold consisting of all real-valued functions in $L^2(0,\infty)$ over $\mathbb{R}$. Define α_t on the Weyl operators by

$$\alpha_t(W_{f+ig}) = W_{(S_t f + iT_t g)}.$$

Using a careful analysis of the generators of the semigroups T_t (related to the analysis carried out above in Lecture V (see Theorem 5.9)) Izumi shows that the corresponding flows are either of type I or III and that the type III examples contain all of the examples described by Tsirelson in [T]. It would be interesting to know if Powers' original example of a type III E_0-semigroup is contained among those in [Iz]. In any case, Izumi's work may be considered a breakthrough in the understanding of the structure of E_0-semigroups of type III.

In order to give a hint of current directions in the field we end these notes by mentioning a couple of problems of great interest to researchers. The first is the development of a comparison theory for E-semigroups, which is at least superficially similar to the comparison of projections in a von Neumann algebra. We require the following definition, due to Powers.

Definition 6.9. *Let α and β be E-semigroups. Then we say that α is a cocycle extension of β (written $\alpha \succcurlyeq \beta$) if β is cocycle conjugate to a subordinate of α.*

Problem 6.1. Suppose $\alpha \succcurlyeq \beta$ and $\beta \succcurlyeq \alpha$. Are α and β then cocycle conjugate?

This problem seems to be difficult in general but some partial results have been obtained. The answer is known to be affirmative when α and β are of type II_0 and arise from (one-dimensional) unbounded boundary weights as we have described above. In this situation one can use Powers' technology of boundary weights along with an analogue of von Neumann's approach to the comparison theory of projections to implement the cocycle conjugacy. This is (as yet unpublished) work of A. Alevras, Powers and Price.

Although a complete classification of E_0-semigroups up to cocycle conjugacy seems to be intractable, Powers has proposed a more modest goal. To describe this, consider an E_0-semigroup α. A *maximal chain* for α is a sequence of E-semigroup subordinates

$$\alpha \succcurlyeq \alpha_1 \succcurlyeq \alpha_2 \succcurlyeq \cdots$$

where one can not add any additional E-semigroups to the sequence, i.e., if $\alpha_i \succcurlyeq \beta \succcurlyeq \alpha_{i+1}$ for any i then either $\beta = \alpha_i$ or $\beta = \alpha_{i+1}$. To specify the length of a maximal chain we could add another subscript to the notation already used to describe an E_0-semigroup. For example, an E_0-semigroup of type $II_{0,0}$ is of type II_0 with a maximal chain of the form $\alpha \succcurlyeq \alpha_1$ where α_1 must be an E-semigroup of type I_0. An E_0-semigroup of type $II_{0,1}$ is one with a maximal chain of the form $\alpha \succcurlyeq \alpha_1 \succcurlyeq \alpha_2$, where α_1 is cocycle conjugate to an E_0-semigroup of type $II_{0,0}$ and α_2 is cocycle conjugate to an E_0-semigroup of type I_0, i.e., to a one-parameter group of automorphisms.

Problem 6.2. Classify all of the cocycle conjugacy classes of the E_0-semigroups of type $II_{m,n}$ where m and n are small.

References

[APP] A. Alevras, R. T. Powers and G. Price, Cocycles for one-parameter flows of $\mathcal{B}(\mathcal{H})$, *J. Func. Anal.* **230**, 1–64.
[A1] W. Arveson, An addition formula for the index of semigroups of endomorphisms of $\mathcal{B}(\mathcal{H})$, *Pacific J. Math.* **137** (1989) 19–36.
[A2] W. Arveson, Continuous analogues of Fock space, *Mem. Amer. Math. Soc.* **80** (1989).
[A3] W. Arveson, Continuous analogues of Fock space II: the spectral C^*-algebra, *J. Func. Anal.* **90** (1990) 138–205
[A4] W. Arveson, *An invitation to C^*-algebras*, Springer (1976).
[A5] W. Arveson, *Noncommutative dynamics and E-semigroups*, Springer (2003).
[A6] W. Arveson, *On the existence of E_0-semigroups*, arXiv:math/0605215.
[A7] W. Arveson, On the index and dilations of completely positive semigroups, *Int. J. Math.* **10** (1999) 791–823.
[Bh1] B. V. R. Bhat, Cocycles of CCR flows, *Mem. Amer. Math. Soc.* **149** (2001).

[Bh2] B. V. R. Bhat, An index theory for quantum dynamical semigroups, *Trans. Amer. Math. Soc.* **348** (1996) 561–583.

[Bh3] B. V. R. Bhat, Minimal dilations of quantum dynamical semigroups to semigroups of endomorphisms of C^*-algebras, *J. Ramanujan Math. Soc.* **14** (1999) 109–124.

[BhS] B. V. R. Bhat and R. Srinivason, *On product systems arising from sum systems* arXiv:math/0405276.

[Br1] O. Bratteli, Derivations, dissipations, and group actions on C^*-algebras, *Lecture Notes in Mathematics* **1229**, Springer (1986).

[Br2] O. Bratteli, Inductive limits of finite-dimensional C^*-algebras, *Trans. Amer. Math. Soc.* **171** (1972) 195–234.

[BR] O. Bratteli and D. W. Robinson, *Operator algebras and quantum statistical mechanics I*, Springer (1979).

[Fi] P. A. Fillmore, *A user's guide to operator algebras*, Wiley (1996).

[Iz] M. Izumi, A perturbation problem for the shift semigroup, *J. Func. Anal.* (to appear).

[IzS] M. Izumi and R. Srinivasan, *Generalized CCR-flows*, arXiv:math07053280.

[K] R. V. Kadison, Derivations on operator algebras, *Ann. Math.* **83** (1966) 280–293.

[Po1] R. T. Powers, Construction of E_0-semigroups from CP-flows, *Amer. Math. Soc. Contemp. Math.* **335** (2003) 57–98.

[Po2] R. T. Powers, Continuous spatial semigroups of completely positive maps of $\mathcal{B}(\mathcal{H})$, *New York J. Math.* **9** (2003) 165–269.

[Po3] R. T. Powers, Fermi field algebra, *Cargèse lectures in physics* **4** (1970) 363–367.

[Po4] R. T. Powers, An index theory for semigroups of $*$-endomorphisms of $\mathcal{B}(\mathcal{H})$ and type II_1 factors, *Canadian J. Math.* **40** (1988) 86–114.

[Po5] R. T. Powers, Induction of semigroups of endomorphisms of $\mathcal{B}(\mathcal{H})$ from completely positive semigroups of $n \times n$ matrix algebras, *Int. J. Math.* **10** (1999) 773–790.

[Po6] R. T. Powers, *A non-spatial continuous semigroup of $*$-endomorphisms of $\mathcal{B}(\mathcal{H})$*, Publ. RIMS (Kyoto University) **23** (1987) 1054–1069.

[Po7] R. T. Powers, New examples of continuous spatial semigroups of endomorphisms of $\mathcal{B}(\mathcal{H})$, *Int. J. Math.* **10** (1999) 215-288.

[Po8] R. T. Powers, A nonspatial continuous semigroup of $*$-endomorphisms of $\mathcal{B}(\mathcal{H})$, *Publ. Res. Inst. Math. Sci.* **23** (1987) 1053–1069.

[PP] R. T. Powers and G. L. Price, Continuous spatial semigroups of $*$-endomorphisms of $\mathcal{B}(\mathcal{H})$, *Trans. Amer. Math. Soc.* (1990) **321** 347–361.

[PR] R. T. Powers and D. W. Robinson, An index for continuous semigroups of endomorphisms of $\mathcal{B}(\mathcal{H})$, *J. Func. Anal.* **84** (1989) 85–96.

[PS] R. T. Powers and E. Størmer, Free states of the canonical anticommutation relations, *Comm. Math. Phys.* **16** (1970) 1–33.

[RSi] M. Reed and B. Simon, Methods of Modern Mathematical Physics, vol. I and II, Academic Press (1972).

[RSz] F. Riesz and B. Sz.-Nagy, *Functional Analysis*, Frederick Ungar Publishing (1955).

[R] W. Rudin, *Functional Analysis*, McGraw Hill (1973).

[S1] S. Sakai, Derivations of simple C^*-algebras, *J. Func. Anal.* **2** (1968) 202–206.

[S2] S. Sakai, Derivations of W^*-algebras, *Ann. Math.* **83** (1966) 273–278.

[S3] S. Sakai, Operator algebras in dynamical systems, Cambridge U. Press (1991).

[SW] I. Singer and J. Wermer, Derivations on commutative normed algebras, *Math. Ann.* **129** (1955) 260–264.

[Sk] M. Skeide, A simple proof of the fundamental theorem about Arveson systems, *Inf. Dim. Anal. Quantum Prob.* (to appear).

[Sti] W. T. Stinespring, Positive functions on C^*-algebras, *Proc. Amer. Math. Soc.* **6** (1955) 211–216.

[Sto] M. H. Stone, *Linear transformations in Hilbert space*, vol. 15, Amer. Math. Soc. Colloquium Publ. (1932).

[T] B. Tsirelson, Non-isomorphic product systems, *Amer. Math. Soc. Contemp. Math.* **335** (2003) 273–328.

[Y] K. Yosida, *Functional Analysis*, Springer-Verlag (1974).

[Za] J. Zacharias, Continuous tensor products and Arveson's spectral C^*-algebras, *Mem. Amer. Math. Soc.* **143** (2000).

Proc. Int. Conf. – Lectures on Operator Theory No. 8, 2009, pp. 87–100.

An Introduction to Basic Functional Analysis

T. S. S. R. K. Rao

Stat–Math Unit, Indian Statistical Institute, R. V. College P.O., Bangalore 560 059, India
e-mail: tss@isibang.ac.in

1. Introduction

This is a write-up of the lectures given at the Advanced Instructional School (AIS) on 'Functional Analysis and Harmonic Analysis' during the first week of July 2006. All of the material is standard and can be found in many basic text books on Functional Analysis.

As the prerequisites, I take the liberty of assuming 1) Basic Metric space theory, including completeness and the Baire category theorem. 2) Basic Lebesgue measure theory and some abstract (σ-finite) measure theory including the Radon-Nikodym theorem and the completeness of L^p-spaces.

2. Banach spaces and examples

Let X be a vector space over the real or complex scalar field. Let $\|\cdot\| : X \to R^+$ be a function such that

(1) $\|x\| = 0 \Leftrightarrow x = 0$
(2) $\|\alpha x\| = |\alpha|\|x\|$
(3) $\|x + y\| \leq \|x\| + \|y\|$

for all $x, y \in X$ and scalars α.

Such a function is called a norm on X and $(X, \|\cdot\|)$ is called a normed linear space. Most often we will be working with only one specific $\|\cdot\|$ on any given vector space X thus we omit writing $\|\cdot\|$ and simply say that X is a normed linear space.

It is an easy exercise to show that $d(x, y) = \|x - y\|$ defines a metric on X and thus there is an associated notion of topology and convergence.

If X is a normed linear space and $Y \subset X$ is a subspace then by restricting the norm to Y, we can consider Y as a normed linear space.

Clearly $|\|x\| - \|y\|| \leq \|x - y\|$ and hence for any sequence $\{x_n\}_{n\geq 1} \subset X$, $x_n \to x \Rightarrow \|x_n\| \to \|x\|$.

When the metric d is complete, i.e., every Cauchy sequence in X converges in X, we say that X is a Banach space.

It is easy to show that the Euclidian spaces R^n and C^n are Banach spaces.

A subspace $Y \subset X$ is a Banach space if and only if it is a closed subset of X.

Example 1. Let K be a compact Hausdorff space, let $C(K)$ denote the vector space of complex-valued continuous functions on K. For any $f \in C(K)$, define $\|f\| = \sup\{|f(k)| : k \in K\}$. This is called the supremum norm and $C(K)$ is a Banach space.

More concretely, let $\Delta = \{z \in C : |z| \leq 1\}$ and $\Gamma = \{z : |z| = 1\}$. Let $A = \{f \in C(\Delta) : f \textit{ is analytic in } \Delta^o\}$. It is easy to see that A is a closed subspace and hence a Banach space. On the other hand $\mathcal{P} = \{p \in C(\Delta) : p \textit{ is a polynomial in } z\}$ is not a closed set and hence is not a Banach space.

If one only considers a locally compact set K then important spaces associated with this are, $C_0(K) = \{f : K \to C : f \textit{ is continuous and vanishes at infinity}\}$ (recall that f vanishes at infinity if for every $\epsilon > 0$ there is a compact set $C \subset K$ such that $|f| < \epsilon$ outside C) and $C_c(K) = \{f : K \to C : f \textit{ is continuous and compactly supported}\}$ (recall that f is compactly supported if $\{k : f(k) \neq 0\}^-$ is a compact set). Both these spaces are equipped with the supremum norm.

It is easy to see that $C_0(K)$ is a Banach space and is the completion (in the metric space sense) of the space $C_c(K)$.

Example 2. Let $(\Omega, \mathcal{A}, \mu)$ be a σ-finite measure space (more concretely take $\Omega = R, \mathcal{A}$ to be the σ-field of Lebesgue measurable sets, μ to be the Lebesgue measure). In what follows let us take two measurable functions that are equal almost every where (a.e) with respect to the measure μ as 'equal'. For $1 \leq p < \infty$, let $L^p(\mu) = \{f : \Omega \to C, f \textit{ is measurable and } \int |f|^p \, d\mu < \infty\}$. For $f \in L^p(\mu)$, let $\|f\| = \left(\int |f|^p \, d\mu\right)^{\frac{1}{p}}$. Then $L^p(\mu)$ is a Banach space.

A measurable function f is said to be essentially bounded if there exist a $\alpha > 0$ such that $\mu((|f|)^{-1}((\alpha, \infty]) = 0$. α is called an essential bound.

Let $L^\infty(\mu) = \{f : \Omega \to C, f \textit{ is measurable and essentially bounded}\}$. For $f \in L^\infty(\mu)$ define $\|f\| = \inf\{\alpha : \alpha \textit{ is an essential bound}\}$. Then $L^\infty(\mu)$ is a Banach space.

A specific situation of the above example and an easy generalization leads to a procedure for generating new classes of Banach spaces. We will only consider countable index sets.

Example 3. Let $\{X_n\}_{n\geq 1}$ be a sequence of Banach spaces. Let $1 \leq p < \infty$. Let $X = \left\{x = \{x_n\} \in \prod X_n : \sum_1^\infty \|x_n\|^p < \infty\right\}$. For $x \in X$, define $\|x\| = \left(\sum_1^\infty \|x_n\|^p\right)^{\frac{1}{p}}$. Then X is a Banach space. Next let $X = \{x = \{x_n\} : \sup |x_n| < \infty\}$. Define $\|x\| = \sup |x_n|$. This is again called the supremum norm. This is again a Banach space.

When all the X_n are taken as the scalar field, the corresponding space X is denoted by ℓ^p and ℓ^∞ respectively.

Other important sequence spaces that are Banach spaces are, c, the space of convergent sequences of scalars and its subspace c_0 of sequences converging to 0. Both the spaces are equipped with the supremum norm. They can also be treated as special cases of Example 1 for the discrete space N and its one point compactification.

Next example again gives a way of generating new Banach spaces from a given collection.

Example 4. Let X be a Banach space, let $Y \subset X$ be a closed subspace. Consider the quotient vector space $X|Y$. Elements of this space are denoted by $[x] = \{x+y : y \in Y\}$. $\pi : X \to X|Y$ defined by $\pi(x) = [x]$ is called the canonical quotient map.

Define $\|[x]\| = d(x, Y) = \inf\{\|x + y\| : y \in Y\}$. Now $X|Y$ is a Banach space.

3. Operators on normed linear spaces, Hahn-Banach theorem

Let X be a normed linear space. $X_1 = \{x \in X : \|x\| \leq 1\}$,the ball with center at 0 and radius 1 is called the closed unit ball. By dilating by an $r > 0$ and translating by an $x_0 \in X$ we get the ball $B(x_0, r) = x_0 + rX_1$.

A set $A \subset X$ is said to be bounded if there exists a $r > 0$ such that $A \subset rX_1$. Equivalently $\|x\| \leq r$ for all $x \in A$.

Clearly any compact set and in particular any convergent sequence, is a bounded set.

Let X, Y be normed linear spaces, let $T : X \to Y$ be a linear map. T is said to be bounded if it maps bounded sets to bounded sets. Such a T is called a bounded operator.

Using linearity of T, we have that T is bounded if and only if $T(X_1)$ is a bounded set if and only if there exists an $r > 0$ such that $\|T(x)\| \leq r\|x\|$ for all $x \in X$.

It is easily seen that a linear map is bounded if and only if it is continuous at 0 and then using the continuity of the vector space operations w.r.t the norm, one has that any bounded linear map is continuous.

Let X be a normed linear space and let $Y \subset X$ be a closed subspace. Let $\pi : X \to X|Y$ be the quotient map. Then $\|\pi\| = 1$.

The following remarks which are like starred exercises, are designed to illustrate the role of finite dimensional spaces.

For the Euclidean spaces, any linear map $T : C^n \to C^m$ is continuous.

A linear, one-to-one and onto map $T : X \to Y$ is said to be an isomorphism if both T and T^{-1} are bounded.

T is an isomorphism if and only if there exists constants $m, M > 0$ such that $m\|x\| \leq \|T(x)\| \leq M\|x\|$.

When $m = M = 1$ we say that T is an isometry.

Any finite dimensional normed linear space is isomorphic to some Euclidean space. So it is easy to see that any finite dimensional normed linear space is a Banach space.

Let X be a normed linear space and $Y \subset X$ a finite dimensional space. Then Y is closed.

We denote by $\mathcal{L}(X, Y)$ the vector space of bounded linear operators.

For $T \in \mathcal{L}(X, Y)$ define $\|T\| = \sup\{\|T(x)\| : x \in X_1\}$. This is a norm. Also for any $x \in X$, $\|T(x)\| \leq \|T\|\|x\|$.

When Y is a Banach space, $\mathcal{L}(X, Y)$ is a Banach space. For $T \in \mathcal{L}(X, Y)$ and $S \in \mathcal{L}(Y, Z)$, $S \circ T \in \mathcal{L}(X, Z)$ and $\|S \circ T\| \leq \|S\|\|T\|$.

When Y is the scalar field, a linear map $f : X \to C$ is called a linear functional. The space of continuous linear functionals is denoted by X^*. For $f \in X^*$, $\|f\| = \sup\{|f(x)| : x \in X_1\}$. X^* is a Banach space.

Let X be an infinite dimensional normed linear space. Using the presence of infinitely many independent unit vectors one can see that there exists a linear map $F : X \to C$ that is not bounded.

Thus it is not clear in an abstract situation how one exhibits non-trivial bounded operators or bounded linear functionals?

Theorem 5 (Hahn-Banach theorem). *Let X be a normed linear space and let $M \subset X$ be a subspace. Let $f \in M^*$. There exists a $g \in X^*$ such that $f = g$ on M and $\|f\| = \|g\|$. In particular for any $x \in X$ there exists an $f \in X^*$ such that $\|x\| = f(x)$.*

Proof. We will assume, by normalizing if necessary, that $\|f\| = 1$.

We shall prove the theorem first when the scalar field is real.

Let $x_0 \notin M$ and let $N = span\{x_0, M\} = \{\alpha x_0 + m : m \in M, \ \alpha \in R\}$.

We will get a $g \in N^*$ satisfying the assertions of the theorem.

Define $g : N \to R$ by $g(\alpha x_0 + m) = \alpha\alpha_0 + f(m)$ where we will choose (below) α_0 to make g bounded by 1. Clearly g is linear and $g = f$ on M.

In order to achieve $\|g\| \leq 1$, we need to show, $|\alpha\alpha_0 + f(m)| \leq \|\alpha x_0 + m\|$ for all α and m.

Dividing by $|\alpha|$, this is same as, $|\alpha_0 + f(\frac{m}{\alpha})| \leq \|x_o + \frac{m}{\alpha}\|$.

Since M is a subspace we only need to achieve $|\alpha_0 + f(m)| \leq \|x_0 + m\|$.

This is equivalent to, $-\|x_0+m\| \leq \alpha_0 + f(m) \leq \|x_0+m\|$ or $-f(m) - \|x_0+m\| \leq \alpha_0 \leq -f(m) + \|x_0 + m\|$.

But for any $x, y \in M$, $|f(x) - f(y)| = |f(x - y)| \leq \|x - y\| \leq \|x + x_0\| + \|x_0 + y\|$. So that $-f(y) - \|x_0 + y\| \leq -f(x) + \|x_0 + x\|$. Thus choose α_0 such that $-f(m) - \|x_0 + m\| \leq \alpha_0 \leq -f(m) + \|x_0 + m\|$. This gives $g \in N^*$ with $\|g\| \leq 1$. As $g = f$ *on* M, $\|g\| = 1$.

To extend f to all of X we next invoke a version of the Axiom of choice called Zorn's Lemma.

Let $\mathcal{G} = \{(N, g) : M \subset N \subset X$ *is a subspace*, $g \in N_1^*, \ g = f$ *on* $M\}$. This is a non-empty set. Define a partial ordering here by 'inclusion' in the first coordinate and functionals agreeing on the smaller space. It is easy to see that every chain here has an upper bound. Thus by the Zorn's Lemma $\mathcal{G}$ has a maximal element, say $((N, g)$. To complete the proof we next note that $N = X$. If not, let $x_0 \notin N$. By the procedure outlined above we get an $(N', g') \in \mathcal{G}$ which is bigger than the maximal element (N, g), leading to a contradiction.

In the case of complex scalar field, we recall that for any linear functional f, its real part *ref* is a real linear functional. Also, given a real linear functional u, $f(x) = u(x) - iu(ix)$ is a complex linear functional whose real part is u.

To see the norm condition, note that $|f(x)| = \alpha f(x)$ for some $\alpha \in \Gamma$. So that $|f(x)| = f(\alpha x) = ref(\alpha x) \leq |ref(\alpha x)| \leq \|\alpha x\| = \|x\|$.

To see the last assertion of the theorem, for $x \in X$, let $M = span\{x\}$. Define $h : M \to C$ by $h(\alpha x) = \alpha\|x\|$. Then $|h(\alpha x)| = |\alpha|\|x\| = \|\alpha x\|$. Thus $\|h\| = 1$ and $h(x) = \|x\|$. By what we have done above, we get $f \in X_1^*$ with $f(x) = \|x\|$. □

Corollary 6. *Let X be a normed linear space and $Y \subset X$ a closed subspace. Let $x_0 \notin Y$ with $\|x_0\| = 1$. There exists an $f \in X^*$ with $f(Y) = 0$ and $f(x_0) = d(x_0, Y)$.*

Proof. Let $Z = span\{Y, x_0\}$. Define $g : Z \to C$ by $g(y + \alpha x_0) = \alpha d(x_0, Y)$. Clearly g is a linear map with $g(Y) = 0$ and $g(x_0) = d(x_0, Y)$. $|g(y + \alpha x_0)| = |\alpha| d(x_0, Y) = d(\alpha x_0, Y) \leq \|\alpha x_0 + y\|$. Thus there is an $f \in X^*$ with the required properties. □

This allows us to associate with a closed subspace $Y \subset X$, $Y^{\perp} = \{f \in X^* : f(Y) = 0\}$. This is called the annihilator of Y and is a closed subspace of X^*.

Let X, Y be normed linear spaces. Now fix a $x_0^* \in X^*$ and $y_0 \in Y$. Define $x_0^* \otimes y_0 : X \to Y$ as $(x_0^* \otimes y_0)(x) = x_0^*(x)y_0$. $x_0^* \otimes y_0$ is an operator with norm $\|x_0^*\|\|y_0\|$ whose range is $span\{y_0\}$. This is called a rank one operator.

The vector space spanned by the set of operators of rank one is the space of finite rank operators and is denoted by $\mathcal{F}$.

The following trick of quotienting an operator is quite useful.

Let $T \in \mathcal{L}(X, Y)$. $\ker(T)$ is a closed subspace of X. Define $T' : X|\ker(T) \to Y$ by $T'([x]) = T(x)$. Then T' is a one-one and continuous operator. If T is onto so is T'.

4. Concrete dual spaces and Banach-Alaoglu theorem

We now identify X^* for some concrete Banach spaces considered in Section 1. Most details can be seen in references [1] and [2].

Let K be a compact Hausdorff space, let $\mathcal{B}$ be the Borel σ-field. For a complex measure μ on $\mathcal{B}$, we recall that the total variation measure $|\mu|$ is defined for $E \in \mathcal{B}$ by $|\mu|(E) = \sup\left\{\sum_1^n |\mu(E_i)| : \{E_i\} \text{ a Borel partition of } E\right\}$. μ is said to be regular if $|\mu|(E) = \inf\{|\mu|(V) : V \text{ open}, E \subset V\}$.

The space of complex regular Borel measures on K with $\|\mu\| = |\mu|(K)$, is a Banach space.

For such a measure μ, $\Lambda : C(X) \to C$ defined by $\Lambda(f) = \int f d\mu$ for $f \in C(K)$, is a bounded linear functional with $\|\Lambda\| \leq \|\mu\|$.

Theorem 7 (Riesz Representation theorem). *Let $\Lambda \in C(X)^*$. There exists a unique regular Borel measure μ such that $\Lambda(f) = \int f d\mu$ for $f \in C(K)$ and $\|\Lambda\| = \|\mu\|$.*

One can see that μ is a probability measure if and only if $\|\mu\| = 1 = \mu(1)$.

There are also versions of the above theorem for locally compact spaces.

Of particular interest are the properties of the Haar Measure μ in the case of a locally compact abelian group.

Now let $1 \leq p \leq \infty$, let q be such that $\frac{1}{p} + \frac{1}{q} = 1$. q is called the conjugate exponent of p. When $p = 1$ we take $q = \infty$.

Let $1 < p \leq \infty$. Let $g \in L^q(\mu)$. Define $\Lambda : L^p(\mu) \to C$ by $\Lambda(f) = \int fg d\mu$ for $f \in L^p(\mu)$. Then $\Lambda \in (L^p(\mu))^*$ and $\|\Lambda\| \leq \|g\|_q$.

Theorem 8. *Let $1 \leq p < \infty$. Let $\Lambda \in (L^p(\mu))^*$. There exists a unique $g \in L^q(\mu)$ such that $\Lambda(f) = \int fg d\mu$ and $\|\Lambda\| = \|g\|_q$.*

We have formulated the above theorem for complex measures, but it is also true for positive σ-finite measures.

For the Haar measure μ, $L^1(\mu)^* = L^\infty(\mu)$ always holds.

Using the discrete version of the above theorem, one can easily identify the dual of the space in Example 3.

Turning to the discussion of the space in Example 4, let $Y \subset X$ be a closed subspace. Define $\Phi : Y^{\perp} \to (X|Y)^*$ by $\Phi(\Lambda)([x]) = \Lambda(x)$. It is easy to see that Φ is an isometry. For $\tau \in (X|Y)^*$ define $\Lambda : X \to C$ by $\Lambda(x) = \tau([x])$. Then $\Lambda \in Y^{\perp}$ and $\Phi(\Lambda) = \tau$.

Next define $\Psi : Y^* \to X^*|Y^\perp$ by $\Psi(\Lambda) = [z^*]$ where $z^* \in X^*$ is a norm preserving extension of Λ given by the Hahn-Banach theorem. Then Ψ is an isometry. For $[x^*] \in X^*|Y^\perp$ define $\Lambda : Y \to C$ by $\Lambda(y) = x^*(y)$. Then $\Lambda \in Y^*$ and $\Psi(\Lambda) = [x^*]$.

These are called the canonical duals of quotient space and subspace.

Consider now X^* as a Banach space. Its dual is denoted by X^{**}. Define $\Psi : X \to X^{**}$ by $\Psi(x)(x^*) = x^*(x)$. Using the Hahn-Banach theorem we have that Ψ is an isometry. This is called the canonical embedding in the bidual. Some times we ignore this embedding and consider $X \subset X^{**}$. A Banach space is said to be reflexive if Ψ is onto.

Clearly any finite dimensional space is reflexive. One has that for $1 < p < \infty$, $L^p(\mu)$ is a reflexive space.

In the infinite dimensional case, there is an example of a $\Lambda \in L^\infty(\mu)^*$ that is not given by a $g \in L^1(\mu)$ so that $\Lambda(f) = \int fgd\mu$.

Thus an infinite dimensional $L^1(\mu)$ and $L^\infty(\mu)$ are not reflexive spaces. The same is true for $C(K)$.

We now discuss briefly a topology on X^* which is weaker than the norm topology and in the case of an infinite dimensional space this topology is not induced by any norm.

Say a net $\{x^*_\alpha\}_{\alpha \in I} \subset X^*$, $x^*_\alpha \to x^* \in X^*$ in the weak*- topology, if for every $x \in X$, $x^*_\alpha(x) \to x^*(x)$.

This defines a topology on X^* and is the smallest topology on X^* w.r.t which $\Psi(X)$ is continuous.

Here weak*-neighborhoods of 0 are of the form $\{x^* : |x^*(x_i)| < \epsilon_i, 1 \leq i \leq n\}$ where $x_i \in X$ and $\epsilon_i > 0$. The vector space operations are continuous in this topology.

Also the identity map $I : (X^*, \|\cdot\|) \to (X^*, weak^*)$ is continuous. We recall that for any normed linear space X, X_1 denotes the closed unit ball.

Theorem 9 (Banach-Alaoglu). *X^*_1 is compact in the weak*-topology. Any weak*-closed and norm bounded set is weak*-compact.*

Proof. Consider the product space $\prod_X \|x\|\Delta$ with the product topology. By the Tychonoff's theorem, this is a compact set. Define $\Phi : (X^*_1, weak^*) \to \prod_X \|x\|\Delta$ by $\Phi(x^*)(x) = x^*(x)$ for $x^* \in X^*_1$ and $x \in X$. Φ is a one-one continuous map.

We now show that $\Phi(X^*_1)$ is a closed set and Φ^{-1} is continuous. Thus Φ is homeomorphism onto its range and as the range is a closed subset of a compact set we conclude that X^*_1 is weak*-compact.

Let $\Phi(x^*_\alpha) \to f \in \prod_X \|x\|\Delta$. Since for any $x \in X$, $f(x) = lim x^*_\alpha(x)$ we have that f is linear and $|f(x)| \leq \|x\|$. Thus $f \in X^*_1$ and $\Phi(f) = f$ so that $\Phi(X^*_1)$ is a closed set.

Also, if $\Phi(x^*_\alpha) \to \Phi(x^*)$ in the product topology, clearly $x^*_\alpha \to x^*$ in the weak*-topology. Therefore Φ is a homeomorphism onto its range. □

Let X be a Banach space, consider X^*_1 equipped with the weak*-topology. Define $\Phi : X \to C(X^*_1)$ defined by $\Phi(x)(x^*) = x^*(x)$. Then Φ is an isometry. Some times we ignore this embedding and consider $X \subset C(X^*_1)$.

Similar to the weak*-topology, one can also consider the smallest topology on X which makes X^* continuous. This is called the weak topology. This is also a vector space topology and the neighborhoods of 0 are of the form $\{x : |x_i^*(x)| < \epsilon_i \ 1 \leq i \leq n\}$ where $x_i^* \in X^*$ and $\epsilon_i > 0$. This topology on X is clearly weaker than the norm topology. Also $x_\alpha \to x$ in weak topology if and only if $x^*(x_\alpha) \to x^*(x)$ for every $x^* \in X^*$. In the case of reflexive spaces the weak and weak*-topology coincide and thus X_1 is a compact set in the weak topology. If X is such that in X^* the weak and weak*-topologies coincide then X is reflexive.

5. Open mapping, closed graph and closed range theorems

In this section, all the spaces are assumed to be Banach spaces. Let X_1^o denote the open unit ball.

Theorem 10 (Open mapping theorem). *Let X, Y be Banach spaces and let $T : X \to Y$ be an onto, bounded operator. There exists a $\delta > 0$ such that $\delta Y_1^o \subset T(X_1^o)$. In particular, T is an open mapping. If T is also one-one then T is an isomorphism.*

Proof. Since $Y = \bigcup_{k \geq 1} T(kX_1^0)$ by the Baire category theorem, there is a $k \geq 1$ and an open set $W \subset Y$ such that $W \subset T(kX_1^o)^-$.

Let $y_0 \in W$ and let $\eta > 0$ such that $y_0 + \eta Y_1^o \subset W$. Thus for any $y \in \eta Y_1^o$ we can choose a sequence $\{x_i\} \subset 2kX_1^o$ with $T(x_i) \to y$.

Now let $\delta = \frac{\eta}{2k}$. Then for any $\epsilon > 0$ and $y \in \delta Y_1^o$ there is a $x \in X_1^o$ with $\|T(x) - y\| < \epsilon$.

Now we use the completeness of X to get a x with $T(x) = y$.

Fix $y \in \delta Y_1^o$. Let $x_1 \in X_1^o$ with $\|y - T(x_1)\| < \frac{1}{2}\delta\epsilon$.

By induction we choose x_{n+1} with $\|x_{n+1}\| < \frac{\epsilon}{2^n}$ and $\|y - \Sigma T(x_i)\| < \frac{\delta\epsilon}{2^{n+1}}$.

Now if $s_n = x_1 + \cdots + x_n$ then it is a Cauchy sequence in X and therefore $s_n \to x \in X$. Since T is continuous, $T(x) = y$ and $\|x\| < 1 + \epsilon$. Since this holds for any $\epsilon > 0$, we have $\delta Y_1^o \subset T(X_1^o)$.

Again by the linearity of T we have that T maps open sets to open sets. □

For a $T : X \to Y$, let $G(T) = \{(x, T(x)) : x \in X\}$. This is called the graph of T. When T is a bounded operator, this is a closed subspace of the product space $X \times Y$.

Theorem 11 (Closed graph theorem). *Let X, Y be Banach spaces and let $T : X \to Y$ be a linear map with $G = G(T)$ closed. Then T is continuous.*

Proof. Make the product space $X \times Y$ a Banach space under the norm $\|(x, y)\| = \|x\| + \|y\|$, so that G is a Banach space. Define $\Phi : G \to X$ by $\Phi(x, T(x)) = x$. This is a linear one-one and onto map. Since projection is continuous w. r. t product topology, we have that Φ is continuous. Therefore, by the open mapping theorem, we have that Φ is an isomorphism.

Thus there exists an $m > 0$ such that $\|x\| + \|T(x)\| = \|(x, T(x))\| \leq m\|x\|$. Thus $\|T(x)\| \leq (m - 1)\|x\|$ and hence T is continuous. □

A linear map $P : X \to X$ is said to be a projection if $P(P(x)) = P(x)$. Note that in this case $X = R(P) \oplus ker(P)$ (direct sum).

Let $Y \subset X$ be a finite dimensional subspace. Then there exists a continuous projection $P : X \to Y$ with $R(P) = Y$. Also if $Y \subset X$ is such that $X|Y$ is finite dimensional the same conclusion holds.

A projection P is continuous if and only if $R(P)$ and $\ker(P)$ are closed.

For any $T \in \mathcal{L}(X, Y)$. Define $T^* : Y^* \to X^*$ by $T^*(y^*)(x) = y^*(T(x))$. T^* is a linear map and $\|T^*(y^*)\| \leq \|T\|\|y^*\|$.

Similarly $T^{**} : X^{**} \to Y^{**}$ is defined and agrees with T on the canonical embedding of X in X^{**}. Thus $\|T\| \leq \|T^{**}\| \leq \|T^*\| \leq \|T\|$. So that $\|T\| = \|T^*\|$.

Note that $T^* : Y^* \to X^*$ is also a continuous map when the domain and the range are equipped with the weak*-topology.

We assume the following consequences of the general version of the Hahn-Banach theorem that we have not proved here.

A subspace $Y \subset X^*$ is weak *-dense if and only if $x \in X$, $y^*(x) = 0$ for all $y^* \in Y$ implies $x = 0$.

Let $A \subset X$ be a closed, convex and balanced set. If $x_0 \notin A$ then there exists an $x^* \in X^*$ such that $|x^*(x)| \leq 1$ for all $x \in A$ but $x^*(x_0) > 1$.

Theorem 12 (Closed Range Theorem). *Let X, Y be Banach spaces. Let $T \in \mathcal{L}(X, Y)$. The range $R(T)$ is closed if and only if $R(T^*)$ is closed if and only if $R(T^*)$ is weak*-closed.*

Proof. Suppose $R(T^*)$ is closed. By replacing Y by $R(T)^-$ if necessary, we may assume that $R(T)$ is dense in Y. Note that this implies that T^* is one-one. Therefore $T^* : Y^* \to R(T^*)$ is an isomorphism by the open mapping theorem. Thus there exists a $c > 0$ such that $c\|y^*\| \leq \|T^*(y^*)\|$ for $y^* \in Y^*$. We next show that $cY_1^0 \subset \overline{T(X_1^0)}$. By an argument similar to the one given during the proof of the Open mapping theorem, this implies that T is onto.

Let $y_0 \in Y$. If $y_0 \notin \overline{T(X_1^0)}$, as the later set is closed, convex and balanced, by the separation theorem quoted above, there exists a $y^* \in Y^*$ such that $|y^*(y)| \leq 1 < |y^*(y_0)|$ for all $y \in \overline{T(X_1^0)}$. For $x \in X_1^0$, $|T^*(y^*)(x)| = |y^*(T(x))| \leq 1$. Thus $c\|y^*\| \leq \|T^*(y^*)\| \leq 1$.

On the other hand $1 < |y^*(y_0)| \leq \|y_0\|\|y^*\| < \frac{1}{c}\|y_0\|$. So $\|y_0\| > c$. Thus $cY_1^0 \subset \overline{T(X_1^0)}$.

Conversely suppose that $R(T)$ is closed. By replacing Y by $R(T)$, we may assume that T is onto. By passing through a quotient if necessary we may assume that T is one-one. Thus T is an isomorphism and hence so is T^*. Therefore $R(T^*)$ is norm as well as weak*-closed. □

Theorem 13 (Uniform boundedness principle). *Let $\{T_n\}_{n\geq 1} \subset \mathcal{L}(X, Y)$. Suppose for every $x \in X$, $\{T_n(x)\}_{n\geq 1}$ is bounded. Then $\{T_n\}_{n\geq 1}$ is bounded.*

Proof. By hypothesis $X = \bigcup_{k\geq 1}\{x : \|T_n(x)\| \leq k \textit{ for all } n\}$. Since X is complete, by the Baire category theorem, there exist k_0, x_0 and $r > 0$ such that $\|x - x_0\| \leq r \Rightarrow \|T_n(x)\| \leq k_o$ for all n. Let M be the bound for $\{T_n(x_0)\}_{n\geq 1}$.

Now for $\|x\| \leq 1$, $\|T_n(x_0 - rx)\| \leq k_o$.

Thus $r\|T_n(x)\| \leq \|T_n(rx - x_0)\| + \|T_n(x_0)\| \leq k_0 + M$.

Hence $\|T_n\| \leq \frac{k_0+M}{r}$. □

6. Theory of operators on Banach spaces

A linear map $T : X \to Y$ is said to be compact if $T(X_1)^-$ is a compact set. We also say that $T(X_1)$ is precompact.

Clearly T is a bounded map. Note that T is compact if and only if every bounded sequence $\{x_n\}_{n\geq 1}$ has a subsequence $\{x_{n_k}\}$ such that $T(x_{n_k}) \to y$ for some $y \in Y$.

It is easy to see that any finite rank operator is compact.

If T is a compact operator with closed range, then T is a finite rank operator.

If $T : X \to Y$ is compact and $Z \subset X$ is a closed subspace then the restriction operator, $T|Z : Z \to Y$ is compact.

If $T : X \to X$ is compact and $\lambda \neq 0$ then $ker(T - \lambda I)$ is finite dimensional.

Let $\mathcal{K}(X, Y)$ denote the vector space of compact operators.

If $T : X \to Y$ is compact and $S : Y \to W$ is bounded operator then $S \circ T$ is compact.

Theorem 14 (Schauder). *T is compact if and only if T^* is compact.*

Proof. Suppose T is compact. Note that since T^* is weak*-continuous, $T^*(Y_1^*)$ is weak*-compact and hence norm-closed. To show that $T^*(Y_1^*)$ is compact, let $\{y_n^*\}_{n\geq 1} \subset Y_1^*$. Since $y_n^* \in C(T(X_1)^-)$ are uniformly bounded and linear maps, it is easy to see that $\{y_n^*\}_{n\geq 1}$ is equicontinuous. Now using the Ascoli's theorem, we can conclude that there is a subsequence $y_{n_k}^*$ converging uniformly on $T(X_1)$.

Further, $\|T^*(y_{n_k}^*) - T^*(y_{n_l}^*)\| = \sup\{|(y_{n_k}^* - y_{n_l}^*)(y)| : y \in T(X_1)\} \to 0$. Since X^* is complete, we get that $T^*(y_{n_k}^*)$ converges. Therefore T^* is compact.

Now suppose T^* is compact. By what we have proved above, T^{**} is compact. Since $T = T^{**}|X$ we have that T is compact. □

When T is compact let us note that for any net, $\{y_\alpha^*\} \subset Y_1^*$, $y_\alpha^* \to y^*$ in the weak*-topology implies, $T^*(y_\alpha^*) \to T^*(y^*)$ in the norm topology (i.e., T^* is completely continuous).

Theorem 15. *$\mathcal{K}(X, Y)$ is a closed subspace of $\mathcal{L}(X, Y)$ and hence is a Banach space.*

Proof. Let $\{T_n\}_{n\geq 1} \subset \mathcal{K}(X, Y)$ and $T_n \to T$. We shall show that $T(X_1)$ is totally bounded. Then, as Y is a Banach space, we have that $T(X_1)^-$ is a compact set. Let $\epsilon > 0$ and let n_0 be such that $\|T_{n_0} - T\| < \epsilon$. Since $T_{n_0}(X_1)$ is a totally bounded set, it is now easy to see that $T(X_1)$ can be covered by finitely many balls with radius 3ϵ. Thus $T \in \mathcal{K}(X, Y)$. □

Thus, if $\mathcal{F}(X, Y) \subset \mathcal{K}(X, Y)$ denote the finite rank operators then $\mathcal{F}(X, Y)^- \subset \mathcal{K}(X, Y)$. From now on we take $X = Y$ so that, $\mathcal{L}(X)$ is an algebra with I as identity and, $\mathcal{K}(X)$ is a two-sided ideal. When X is infinite dimensional, $I \notin \mathcal{K}(X)$. For any $T \in \mathcal{L}(X)$, its spectrum $\sigma(T) = \{\lambda : T - \lambda I \textit{ is not invertible}\}$. λ is said to be an eigenvalue if $0 \neq x \in \ker(T - \lambda I)$. When X is infinite dimensional and T compact, $0 \in \sigma(T)$.

Proposition 16. *For a compact operator T and a non-zero eigenvalue λ, $R(T - \lambda I) \neq X$. $E_r = \{\lambda : |\lambda| > r, \lambda$ an eigenvalue$\}$ is a finite set and thus E, the set of eigenvalues, is a countable set.*

Proof. Suppose the conclusions fail. We claim that there exists a sequence $\{M_n\}_{n\geq 1}$ of strictly increasing closed subspaces of X with $T(M_n) \subset M_n$ and $(T - \lambda_n I)(M_n) \subset M_{n-1}$, where $\{\lambda_n\}_{n\geq 1} \subset E$ is bounded below by an $r > 0$.

Assuming the claim, as M_{n-1} is a proper subspace of M_n, let $y_n \in M_n$ with $\|y_n\| \leq 2$ and $d(y_n, M_{n-1}) = 1$.

For $2 \leq m < n$, let $z = T(y_m) - (T - \lambda_n I)(y_n)$. Then, $\|T(y_n - y_m)\| = \|\lambda_n y_n - z\| = |\lambda_n|\|y_n - \frac{z}{\lambda_n}\| \geq |\lambda_n| > r$. This contradicts the compactness of T.

Thus, suppose $range(S = (T - \lambda I)) = X$. Let $M_n = Ker(S^n)$. Since λ is an eigenvalue, there exists a $0 \neq x_1 \in M_1$ and as $range(S) = X$, there exists a sequence $\{x_n\}_{n\geq 1} \subset X$ such that $S(x_{n+1}) = x_n$. Now $S^n(x_{n+1}) = x_1 \neq 0$ but $S^{n+1}(x_{n+1}) = S(x_1) = 0$. Thus M_n is a proper closed subspace of M_{n+1}. The other assertions are valid with $\lambda_n = \lambda$.

On the other hand, suppose for an $r > 0$ there is a distinct sequence $\{\lambda_n\}_{n\geq 1} \subset E$ with $|\lambda_n| > r$. Let $\{e_n\}_{n\geq 1}$ be a corresponding sequence of independent eigenvectors. Let $M_n = span\{e_i\}_{1\leq i\leq n}$. Then $T(M_n) \subset M_n$ and $(T - \lambda_n I)(M_n) \subset M_{n-1}$. □

We recall that for an operator T, $R(T)$ stands for its range.

Theorem 17. *Let* $T \in \mathcal{K}(X)$, $\lambda \neq 0$. *The numbers,* $\dim(\ker(T - \lambda I)) = \dim(X|R(T - \lambda I)) = \dim(\ker(T^* - \lambda I))$. *If* $0 \neq \lambda \in \sigma(T)$ *then* λ *is an eigenvalue of* T *and* T^*. $\sigma(T)$ *is countable, compact with at most one limit point,* 0.

Proof. Let $S = T - \lambda I$ then $S^* = T^* - \lambda I$. We will first show that $R(S)$ is closed. Since $\ker(S)$ is finite dimensional, let $P : X \to \ker(S)$ be a bounded projection. It is enough to show that S is bounded below on $\ker(P)$ where it is one-one. If not, there exists a sequence $\{x_n\}_{n\geq 1} \subset \ker(P)$ of unit vectors such that $S(x_n) \to 0$. Since T is compact we may assume that $x_n \to x_0 \in \ker(P)$, so that $\lambda x_n \to \lambda x_o$. Since S is one-one on $\ker(P)$, $S(x_0) = 0 \Rightarrow x_0 = 0$. This contradicts that x_n's are unit vectors and $\lambda \neq 0$. Now by duality, $\ker S^* = R(S)^\perp = (X|R(S))^*$, so that $\dim(X|R(T - \lambda I)) = \dim(\ker(T^* - \lambda I))$.

Now suppose $\dim(\ker(S)) > \dim(X|R(S))$. Suppose $X = \ker(S) \oplus E = F \oplus R(S)$. Let P be the projection with range $\ker(S)$.

Since $\dim(\ker(S)) > \dim(F)$, let $\phi : \ker(S) \to F$ be linear onto and $\phi(x_0) = 0$ for some $x_0 \neq 0$. Let $\Phi = T + \phi \circ P$. Clearly Φ is compact and $\Phi - \lambda I = S + \phi \circ P$. As $\Phi(x_0) = \lambda x_0$, $R(\Phi - \lambda I) \neq X$. Since $P = 0$ on E, $(\Phi - \lambda I)(E) = R(S)$. Also as $P = I$ on $\ker(S)$, $(\Phi - \lambda I)(\ker(S) = F$. Therefore, $R(\Phi - \lambda I) = X$, a contradiction.

Thus $\dim(\ker(T - \lambda I)) \leq \dim(X|R(T - \lambda I))$ and by duality we have equality.

We already know that $\sigma(T) \cup \{0\}$ is countable and compact. If X is finite dimensional, $\sigma(T)$ is a finite set and if X is infinite dimensional $0 \in \sigma(T)$. □

7. Hilbert spaces

Let H be a complex linear space. $\langle , \rangle : H \times H \to C$ such that, $\langle , \rangle$ is linear in the first variable with $\langle x, x\rangle \geq 0$ and $\langle x, x\rangle = 0$ if and only if $x = 0$, $\langle x, y\rangle = \langle y, x\rangle^-$, is called an inner product on H.

For a σ-finite measure space $(\Omega, \mathcal{A}, \mu)$ and $p = 2$, for $f, g \in L^2(\mu)$, $\langle f, g\rangle = \int fg^- d\mu$ is an inner product.

Similarly, in the discrete case ℓ^2 the space of square summable complex sequences, $\langle x, y\rangle = \sum x(n)y(n)^-$ is an inner product.

If $\langle x, y\rangle = 0$ we say, x is orthogonal to y, and write $x \perp y$.

Let $\|x\| = (\langle x, x\rangle)^{\frac{1}{2}}$. Put $\alpha = \langle x, y\rangle$.

For any $\lambda \in C$, $0 \leq \|\lambda x + y\|^2 = |\lambda|^2\|x\|^2 + 2re(\alpha\lambda) + \|y\|^2$.

Thus, if $\langle x, y\rangle = 0$ then $\|y\| \leq \|\lambda x + y\|$ for all $\lambda \in C$.

Also when $0 \neq x$, by taking $\lambda = -\frac{\alpha^-}{\|x\|^2}$ we see that, $0 \leq \|\lambda x + y\| = \|y\|^2 - \frac{|\alpha|^2}{\|x\|^2}$.

Thus, $|\langle x, y\rangle| \leq \|x\|\|y\|$. This is called the Schwarz inequality.

Also if $\langle x, y\rangle \neq 0$ for the above choice of λ, $\|y\| > \|\lambda x + y\|$. Hence $x \perp y$ if and only if $\|y\| \leq \|\lambda x + y\|$ for all $\lambda \in C$.

Thus orthogonality can be described only using $\|\cdot\|$. In a Banach space, according to G. Birkhoff, $x \perp y$ if $\|\lambda x + y\| \geq \|y\|$ for all $\lambda \in C$.

It is easy to see now from the Schwarz inequality that $\|x + y\| \leq \|x\| + \|y\|$. Thus $\|x\| = (\langle x, x\rangle)^{\frac{1}{2}}$ defines a norm on H.

H is a Hilbert space when it is a Banach space with this norm.

$L^2(\mu)$ and ℓ^2 are Hilbert spaces with norms, $\|f\|_2 = \left(\int |f|^2 d\mu\right)^{\frac{1}{2}}$ and $\|x\|^2 = \left(\sum |x(n)|^2\right)^{\frac{1}{2}}$.

The equation $\|x+y\|^2 + \|x-y\|^2 = 2(\|x\|^2 + \|y\|^2)$ is called the the parallelogram law.

For $M \subset H$ define $M^\perp = \{x \in H : \langle x, y\rangle = 0 \; for \; y \in M\}$. This is a closed subspace. Also $M \cap M^\perp = 0$.

Fix a $y \in H$. Define $f : H \to C$ by $f(x) = \langle x, y\rangle$. Then $f \in H^*$ and $\|f\| = \|y\|$.

Theorem 18. *Every non-empty closed convex set $A \subset H$ has a unique vector of smallest norm. For any closed subspace $M \subset H$ there is a unique projection $P \in \mathcal{L}(H)$ such that* $\ker(P) = M^\perp$, $R(P) = M$ *and* $\|x\|^2 = \|P(x)\|^2 + \|(I - P)(x)\|^2$. *Let $f \in H^*$. There exists a unique $y \in H$ such that $f(x) = \langle x, y\rangle$.*

Proof. We may assume that $0 \notin A$. Let $d = d(0, A)$. Let $\{a_n\}_{n\geq 1} \subset A$ and $d = \lim \|a_n\|$. As $\frac{1}{2}(a_n + a_m) \in A$, $\|a_n + a_m\|^2 \geq 4d^2$. Thus by the Parallelogram law, $\|a_n - a_m\|^2 \to 0$. therefore $a_n \to a \in A$ and hence $d = \|a\|$. If $b \in A$ and $\|b\| = d$ then the sequence $a_n = a \; or \; b$ alternately must converge by the above argument. So $a = b$.

Let $x \in H$ and let $d(x, M) = \|x - y\|$ for a unique $y \in M$. Define $P(x) = y$. Now $\|x - P(x)\| \leq \|x - P(x) + m\|$ for all $m \in M$. Therefore $x - P(x) \in M^\perp$. Thus P is a projection with the above properties.

Let $0 \neq f \in H^*$. Applying the above conclusion to $M = \ker(f)$, let $0 \neq z \in M^\perp$. Since $f(x)z - f(z)x \in \ker(f)$, $f(x)\langle z, z\rangle = f(z)\langle x, z\rangle$. Now by taking $y = \frac{f(z)^-}{\langle z,z\rangle} z$ we have $f(x) = \langle x, y\rangle$. Uniqueness is easy to see. □

The P in the above theorem is called an orthogonal projection.

Note that $f \to y$ is a conjugate linear isometry of H^* with H. Thus a Hilbert space is reflexive and the unit ball is weakly compact.

We can relate $M^\perp$ to the annihilator of M discussed in Section 3.

We saw, in the case of Banach spaces, $\sum_{n=1}^{\infty} \|x_n\| < \infty$ implies $\sum_{n=1}^{\infty} x_n = \lim \sum_{i=1}^{n} x_i$ exists. This property, in the case of pair-wise orthogonal vectors, can be completely described in terms of the norm in a Hilbert space.

Theorem 19. *Let $\{x_n\}_{n \geq 1} \subset H$ be a sequence of pair-wise orthogonal vectors. $\sum x_n$ converges if and only if $\sum \|x_n\|^2 < \infty$. This is also equivalent to $\sum \langle x_n, y \rangle < \infty$ for any $y \in H$.*

Proof. As the vectors are pair-wise orthogonal, by the parallelogram law we have, $\sum_{i=k}^{l} \|x_i\|^2 = \|x_k + \cdots + x_l\|^2$. Thus $\sum \|x_n\|^2 < \infty$ implies the convergence of the series of vectors. Also by the Schwarz' inequality, $\sum x_n$ converges, implies $\sum \langle x_n, y \rangle < \infty$ for all $y \in H$.

Conversely, suppose $\sum \langle x_n, y \rangle < \infty$ for all $y \in H$. Define $\Lambda_n : H \to C$ by $\Lambda_n(x) = \sum_{i=1}^{n} \langle x_i, y \rangle$. Then $\Lambda_n \in H^*$ and $\|\Lambda_n\| = \sum_{i=1}^{n} \|x_i\|^2$. Also for any $y \in H$, $\{\Lambda_n(y)\}$ is a bounded sequence. Thus, by the uniform boundedness principle, we get that $\sum \|x_n\|^2 < \infty$. □

In the following we assume that H is a separable space.

A sequence $\{x_n\}_{n \geq 1}$ of unit vectors that are pair-wise orthogonal is called an orthonormal sequence. These are clearly independent vectors. There can be only countably many such vectors in a separable space.

Note that in the case of ℓ^2, $x_n = e_n$ the coordinate vectors form an orthonormal sequence and $\|x\|^2 = \sum |\langle x, e_i \rangle|^2$ and $\langle x, y \rangle = \sum \langle x, e_n \rangle \langle e_n, y \rangle$.

An orthonormal sequence satisfies the Bessel's inequality, $\sum |\langle x, x_n \rangle|^2 \leq \|x\|^2$.

To see this, let $M_n = span\{\{x_i\}_{1 \leq i \leq n}\}$. Then $d(x, M) = \|x - \sum \langle x, x_i \rangle x_i\|$.

Also, we know $\|x\|^2 = \|x - \sum \langle x, x_i \rangle x_i\|^2 + \|\sum \langle x, x_i \rangle x_i\|^2$.

But $\|\sum \langle x, x_i \rangle x_i\|^2 = \sum |\langle x, x_i \rangle|^2$.

Thus $\Phi : H \to \ell^2$ defined by $\Phi(x) = \{\langle x, x_n \rangle\}$ is a linear onto map with $\|\Phi\| = 1$.

We will next see conditions under which Φ is an isometry and preserves the inner product.

An orthonormal sequence is said to be maximal if $\langle x, x_n \rangle = 0$ for all n implies $x = 0$.

One can show that $\{x_n\}_{n \geq 1}$ is a maximal orthonormal sequence if and only if $M = span\{x_n\}_{n \geq 1}$ is dense in H.

Theorem 20. *$\{x_n\}_{n \geq 1}$ is a maximal orthonormal sequence if and only if $\|x\|^2 = \sum |\langle x, x_n \rangle|^2$. In this case, $\langle x, y \rangle = \sum \langle x, x_n \rangle \langle x_n, y \rangle$.*

Proof. Suppose $\{x_n\}_{n \geq 1}$ is a maximal orthonormal sequence. Let $\epsilon > 0$. There is a $z \in M_n$ for some n such that $\|x - z\| < \epsilon$. Therefore $\|x - \sum_1^n \langle x, x_i \rangle x_i\| = d(x, M_n) \leq \|x - z\| < \epsilon$.

Thus, $(\|x\| - \epsilon)^2 < \|\sum_1^n \langle x, x_i \rangle x_i\|^2 = \sum_1^n |\langle x, x_i \rangle|^2 \leq \sum_1^{\infty} |\langle x, x_i \rangle|^2$.

Hence $\|x\|^2 = \sum |\langle x, x_n \rangle|^2$. The converse is trivial. The last assertion follows since $\Phi : H \to \ell^2$ has this property. More precisely, by our hypothesis we have $\langle x, x \rangle = \sum \langle x, x_n \rangle \langle x_n, x \rangle$. For $x, y, \lambda \in C$, $\langle x + \lambda y, x + \lambda y \rangle = \langle \Phi(x) + \lambda \phi(y), \Phi(x) + \lambda \Phi(y) \rangle$. Thus, $\lambda^- \langle x, y \rangle + \lambda \langle y, x \rangle = \lambda^- \langle \Phi(x), \Phi(y) \rangle + \lambda \langle \Phi(y), \Phi(x) \rangle$. Taking $\lambda = 1, i$ in this relation shows that $\langle x, y \rangle = \langle \Phi(x), \Phi(y) \rangle$ as they have the same real and imaginary parts. □

Finally to generate a maximal orthonormal set (also called a complete orthonormal basis) we invoke Zorn's Lemma. Consider the class of all orthonormal sets and order it by inclusion. Every chain here has a maximal element. Thus the maximal element given by the Zorn's Lemma is a maximal orthonormal set.

8. Operators on Hilbert spaces

Let $T \in \mathcal{L}(H)$. If $\langle T(x), x\rangle = 0$ for all $x \in H$ then $T = 0$.

To see this, apply the idea from the last section to get $\langle T(x), y\rangle = 0$ for all $x, y \in H$. Taking $y = T(x)$ thus gives the conclusion.

For each $y \in H$, by Theorem 18, there exists a unique $S(y) \in H$ with $\|S(y)\| \leq \|T(x)\|\|y\|$, such that $\langle T(x), y\rangle = \langle x, S(y)\rangle$. It is easy to see that S is linear and thus $S \in \mathcal{L}(H)$ and $\|S\| \leq \|T\|$. We denote this S by T^* and is unique with the property $\langle T(x), y\rangle = \langle x, T^*(y)\rangle$. It is called the adjoint of T.

Applying the same idea to $T^* \in \mathcal{L}(H)$, $\langle T(x), y\rangle = \langle T^*(y), x\rangle^- = \langle y, T^{**}(x)\rangle^- = \langle T^{**}(x), y\rangle$ so that by uniqueness $T^{**} = T$ and $\|T^*\| = \|T\|$.

Also $\|T(x)\|^2 = \langle T(x), T(x)\rangle = \langle T^*(T(x)), x\rangle \leq \|T^*T\|\|x\|^2$. So $\|T\|^2 \leq \|T^*T\| \leq \|T\|\|T^*\| = \|T\|^2$. Therefore $\|T\|^2 = \|T^*T\|$.

$T \to T^*$ is a conjugate linear map and $(TS)^* = S^*T^*$.

In the case of Hilbert space, we have that T is compact if and only if $T(H_1)$ is a compact set. T is compact if and only if T^* is compact.

For $\alpha \in \ell^\infty$, define $M_\alpha : \ell^2 \to \ell^2$ by $M_\alpha(x) = (\alpha(n)x(n))$. It is easy to see that $\|M_\alpha\| = \|\alpha\|$. One can easily compute M_α^*, and also decide when is M_α compact?

Let H be separable with a complete orthonormal basis $\{e_n\}_{n\geq 1}$. Let P_n be the orthogonal projection onto $span\{e_1, \ldots, e_n\}$. Then $P_n(x) \to x$. If $T \in \mathcal{K}(H)$ then $P_nT \to T$. Thus $\mathcal{F}(H)^- = \mathcal{K}(H)$.

The duality between the range of an operator and the kernel of the adjoint, in Section 5, in the case of Hilbert spaces is given by, $Ker(T^*) = R(T)^\perp$ and $Ker(T) = R(T^*)^\perp$.

T is said to be normal if $TT^* = T^*T$, self-adjoint if $T^* = T$ and unitary if $T^*T = I = TT^*$.

For a normal operator we have, $\|T(x)\|^2 = \langle T(x), T(x)\rangle = \langle T^*T(x), x\rangle = \langle TT^*(x), x\rangle$. Also $\langle TT^*(x), x\rangle = \langle T^*(x), T^*(x)\rangle = \|T^*(x)\|^2$. Thus $\|T(x)\| = \|T^*(x)\|$.

M_α defined above is a normal operator. Define $U : \ell^2 \to \ell^2$ by $U(x_1, \ldots, x_n, \ldots) = (0, x_1, x_2, \ldots)$. It is easy to compute U^*. This is called the shift operator.

We see that an operator U is unitary if and only if U is onto and preserves the norm or inner product.

Let $P \in \mathcal{H}$ be a projection, i.e., $P^2 = P$. Then if P is normal then P is an orthogonal projection (i.e., $\ker(P)^\perp = R(P)$) and P is self adjoint. To see this we note that for a normal operator T, $\ker(T) = \ker(T^*) = R(T)^\perp$. Now since $R(P) = \ker(I - P)$, we have that P is an orthogonal projection.

Now $\|P(x)\|^2 = \langle P(x), P(x)\rangle = \langle P(x), (I - P)(x) + P(x)\rangle$.

Similarly, $\|P(x)\|^2 = \langle P(x) + (I - P)(x), P(x)\rangle$. Thus, $\langle P(x), x\rangle = \langle x, P(x)\rangle$. Hence an orthogonal projection is self adjoint.

References

[1] W. Rudin, *Real and complex analysis*, Second edition, McGraw-Hill Series in Higher Mathematics, McGraw-Hill Book Co., New York-Düsseldorf-Johannesburg (1974).

[2] W. Rudin, *Functional analysis*, McGraw-Hill Series in Higher Mathematics, McGraw-Hill Book Co., New York-Düsseldorf-Johannesburg (1973) xiii+397 pp.

Made in the USA
Monee, IL
05 November 2026

67811367R00061